BIOMATERIAL APPLICATIONS

Macro to Nanoscales

BIOMATERIAL APPLICATIONS

Macro to Nanoscales

Edited by
**Sabu Thomas, PhD, Nandakumar Kalarikkal, PhD,
Weimin Yang, MD, and Snigdha S. Babu**

Apple Academic Press

TORONTO   NEW JERSEY

Apple Academic Press Inc. | Apple Academic Press Inc.
3333 Mistwell Crescent | 9 Spinnaker Way
Oakville, ON L6L 0A2 | Waretown, NJ 08758
Canada | USA

©2015 by Apple Academic Press, Inc.

First issued in paperback 2021

Exclusive worldwide distribution by CRC Press, a member of Taylor & Francis Group
No claim to original U.S. Government works

ISBN 13: 978-1-77463-349-6 (pbk)
ISBN 13: 978-1-77188-027-5 (hbk)

Library of Congress Control Number: 2014953120

Library and Archives Canada Cataloguing in Publication

International Conference on Natural Polymers, Bio-Polymers, Bio-Materials, their Composites, Blends, IPNs, Polyelectrolytes and Gels: Macro to Nano Scales (3rd : 2012 : Kottayam, India)
Biomaterial applications : macro to nanoscales / edited by Sabu Thomas, PhD, Nandakumar Kalarikkal, PhD, Weimin Yang, MD, and Snigdha S. Babu.
Based on papers presented at the ICNP 2012 – Third International Conference on Natural Polymers, Bio-Polymers, Bio-Materials, their Composites, Blends, IPNs, Polyelectrolytes and Gels: Macro to Nano Scales took place at Mahatma Gandhi University, Kottayam, Kerala, India, on October 26, 27, and 28, 2012.

Includes bibliographical references and index.
ISBN 978-1-77188-027-5 (bound)
1. Biopolymers--Congresses. 2. Biomedical materials--Congresses. I. Thomas, Sabu, editor II. Kalarikkal, Nandakumar, editor III. Weimin, Yang, MD, editor IV. Babu, Snigdha S., editor V. Title.

| TP248.65.P62I57 2012 | 572 | C2014-906936-7 |

Apple Academic Press also publishes its books in a variety of electronic formats. Some content that appears in print may not be available in electronic format. For information about Apple Academic Press products, visit our website at **www.appleacademicpress.com** and the CRC Press website at **www.crcpress.com**

ABOUT THE EDITORS

Sabu Thomas, PhD

Dr. Sabu Thomas is the Director International and Inter University Centre for Nanoscience and Nanotechnology Mahatma Gandhi University, Kottayam, India. He is also a full professor in the School of Chemical Sciences at the same University. He is a fellow of many professional bodies. Professor Thomas has authored or co-authored many papers in international peer-reviewed journals in the area of polymer processing. He has organized several international conferences and has more than 420 publications, 11 books and two patents to his credit. He has been involved in a number of books both as author and editor. He is a reviewer to many international journals and has received many awards for his excellent work in polymer processing. His h Index is 67. Professor Thomas is listed as the 5th position in the list of Most Productive Researchers in India, in 2008.

Nandakumar Kalarikkal, PhD

Nandakumar Kalarikkal, PhD, is Associate Professor and Head of Advanced Materials Laboratory, School of Pure and Applied Physics, Mahatma Gandhi University, Kottayam, India. He is also the Joint Director of the International and Inter University Centre for Nanoscience and Nanotechnology at the same university. He is a fellow of many professional bodies. He has authored many professional journal articles and has co-edited several books. He is actively involved in research, and his group works on the synthesis, characterization and applications of various nanomaterials, ion irradiation effects on various novel materials, and phase transitions.

Weimin Yang, MD

Dr. Yang Weimin is the Taishan Scholar Professor of Qingdao University of Science and Technology in China. He is a Professor of the College of Mechanical and Electrical Engineering and Director of the Department of International Exchanges and Cooperation, Beijing University of Chemical

Technology, Beijing , China. In addition, he is a fellow of many professional organizations. Professor Weimin has authored many papers in international peer-reviewed journals in the area of polymer processing. He has contributed to a number of books as author and editor and acts as a reviewer to many international journals. In addition, he is a consultant to many polymer equipment manufacturers. He has also received numerous awards for his work in polymer processing. His interests include polymer processing and CAD/CAE/CAM of polymer processing.

Snigdha S. Babu

Snigdha S. Babu is doing doctoral research in tissue engineering at the International and Inter University Centre for Nanoscience and Nanotechnology at Mahatma Gandhi University, Kottayam, Kerala, India.

CONTENTS

List of Contributors .... *ix*

List of Abbreviations .... *xiii*

Preface .... *xvii*

1. **Green Organic-inorganic Hybrid Material from Plant Oil Polyol** 1
    Eram Sharmin, Mudsser Azam, Fahmina Zafar, Deewan Akram, Qazi Mohd. Rizwanul Haq, and Sharif Ahmad

2. **Bio-Hybrid 3D Tubular Scaffolds for Vascular Tissue Engineering—A Materials Perspective** 15
    Harsh Patel, Roman Garcia, and Vinoy Thomas

3. **Polymers for Use in the Monitoring and Treatment of Waterborne Protozoa** 49
    Helen Bridle and Moushumi Ghosh

4. **Synthesis of Polypyrrole/TiO2 Nanoparticles in Water by Chemical Oxidative Polymerization** 73
    Yang Tan, Michel B. Johnson, and Khashayar Ghandi

5. **Poly (Lactic Acid) Based Hybrid Composite Films Containing Ultrasound Treated Cellulose and Poly (Ethylene Glycol) As Plasticizer and Reaction Media** 101
    Katalin Halász, Mandar P. Badve, and Levente Csóka

6. **An Experimental Observation of Disparity in Mechanical Properties of Turmeric Fiber Reinforced Polyester Composites** 123
    Nadendla Srinivasababu, J. Suresh Kumar and K. Vijaya Kumar Reddy

7. **Wavelength Dependence of SERS Spectra of Pyrene** 135
    F. Hubenthal, D. Blázquez Sánchez, R. Ossig, H. Schmidt, and H.-D. Kronfeldt

8. **Emerging Therapeutic Applications of Bacterial Exopolysaccharides** 145
    P. Priyanka, A. B. Arun, and P. D. Rekha

9. **Preparation and Properties of Composite Films from Modified
    Cellulose Fiber-Reinforced With Different Polymers**..........................169
    Sandeep S. Laxmishwar and G. K. Nagaraja

10. **Natural Bio Resources: The Unending Source of Nanofactory**..........195
    Balaprasad Ankamwar

    Index..*203*

LIST OF CONTRIBUTORS

Sharif Ahmad
Materials Research Laboratory, Department of Chemistry, Jamia Millia Islamia, New Delhi–110025, India

Deewan Akram
Materials Research Laboratory, Department of Chemistry, Jamia Millia Islamia, New Delhi–110025, India. Department of Chemistry, Faculty of Science, Jazan University, PO Box 2097, Jazan, Kingdom of Saudi Arabia

Balaprasad Ankamwar
Bio-Inspired Materials Science Laboratory, Department of Chemistry, University of Pune, Ganeshkhind, Pune–411007, India. E-mail:bankamwar@yahoo.com Phone: +91-20-25601397, Ext.533; Fax: +91-20-25691728

A. B. Arun
Yenepoya Research Centre, Yenepoya University, Deralakatte, Mangalore, Karnataka State, India

Mudsser Azam
Microbiology Research Laboratory, Department of Biosciences, Jamia Millia Islamia, New Delhi–110025, India

Mandar P. Badve
Department of Chemical Engineering, Institute of Chemical Technology, Mumbai, India

Dr. Helen Bridle
Royal Academy of Engineering/EPSRC Research Fellow, Institute of Biological Chemistry, Biophysics and Bioengineering, Heriot-Watt University, Riccarton, Edinburgh, EH14 4AS, Scotland

Levente Csóka
Department of Composite and Paper Technologies, Institute of Wood Based Products and Technologies, University of West Hungary, Sopron, Hungary, E-mail: hakat@sopron.nyme.hu

Khashayar Ghandi
Department of Chemistry and Biochemistry, Mount Allison University, NB, Canada, E4L 1G8 E-mail: kghandi@mta.ca, Tel: 506.961.080

Dr. Moushumi Ghosh
Thapar University, India

Katalin Halász
Department of Composite and Paper Technologies, Institute of Wood Based Products and Technologies, University of West Hungary, Sopron, Hungary, E-mail: hakat@sopron.nyme.hu

Qazi Mohd. Rizwanul Haq
Microbiology Research Laboratory, Department of Biosciences, Jamia Millia Islamia, New Delhi–110025, India

F. Hubenthal
Universität Kassel, Institut für Physik and Center for Interdisciplinary Nanostructure Science and Technology-CINSaT, Heinrich-Plett-Str. 40, 34132 Kassel, Germany, E-mail: hubentha@physik.uni-kassel.de

Michel B. Johnson
Institute for Research in Materials and Department of Chemistry, Dalhousie University, NS, Canada, B3H 4R2

H. D. Kronfeldt
Technische Universität Berlin, Institut für Optik und Atomare Physik, Hardenbergstr. 36, 10623 Berlin, Germany, E-mail: kf@physik.tu-berlin.de

J. Suresh Kumar
Professor, Department of Mechanical Engineering, JNTUH College of Engineering, Kukatpally, Hyderabad-85, A.P., India

Sandeep S. Laxmishwar
Department of studies in Chemistry, Mangalore University, Mangalagangothri, Karnataka-574199, India

G. K. Nagaraja
Department of studies in Chemistry, Mangalore University, Mangalagangothri, Karnataka–574199, India

R. Ossig
Universität Kassel, Institut für Physik and Center for Interdisciplinary Nanostructure Science and Technology-CINSaT, Heinrich-Plett-Str. 40, 34132 Kassel, Germany, E-mail: hubentha@physik.uni-kassel.de

P. Priyanka
Yenepoya Research Centre, Yenepoya University, Deralakatte, Mangalore, Karnataka State, India

K. Vijaya Kumar Reddy
Professor, Department of Mechanical Engineering, JNTUH College of Engineering, Kukatpally, Hyderabad-85, A.P., India

Dr. P. D. Rekha
Deputy Director, Yenepoya Research Center, Yenepoya University, University Road, Deralakatte, Mangalore–575018, Karnataka State, India. Tel: +91 824 2204668; Mobile: +91 9741501821; Fax: +91 824 2204673. Email: rekhapd@hotmail.com, dydirectoryrc@yenepoya.edu.in

D. Blázquez Sánchez
Universität Kassel, Institut für Physik and Center for Interdisciplinary Nanostructure Science and Technology-CINSaT, Heinrich-Plett-Str. 40, 34132 Kassel, Germany, E-mail: hubentha@physik.uni-kassel.de

H. Schmidt
Universität Bayreuth, Forschungsstelle für Nahrungsmittel qualität E. C. Baumann-Str. 20, 95326 Kulmbach, Germany, E-mail: heinar.schmidt@uni-bayreuth.de

Eram Sharmin
Materials Research Laboratory, Department of Chemistry, Jamia Millia Islamia, New Delhi–110025, India. Department of Pharmaceutical Chemistry, College of Pharmacy, Umm Al-Qura University, Makkah Al-Mukarramah, Po Box 715, Postal Code: 21955, Kingdom of Saudi Arabia. E-mail: eram-sharmin@gmail.com

Nadendla Srinivasababu
Professor, Department of Mechanical Engineering, Vignan's Lara Institute of Technology and Science, Vadlamudi-522 213, A. P., India. E-mail: cnjlms22@yahoo.co.in

Yang Tan
Department of Chemistry and Biochemistry, Mount Allison University, NB, Canada, E4L 1G8

Fahmina Zafar
Materials Research Laboratory, Department of Chemistry, Jamia Millia Islamia, New Delhi–110025, India. Inorganic Materials Research Laboratory, Department of Chemistry, Jamia Millia Islamia, New Delhi–110025, India

LIST OF ABBREVIATIONS

AD	Aldehyde
BC	Bacterial Cellulose
BMSC	Bone Marrow Stem Cell
BS	Barbated Skullcup
BT	Biotechnology
CABG	Coronary Artery Bypass Graft
CB	Conduction Band
CNF	Cellulose Nanofibres
CNMB	Center for Nanoscale Materials and Biointegration
CNW	Cellulose Nanowhiskers
COWP1	Cryptosporidium (oo)cysts Wall Protein 1
CT	Chemically Treated
CVD	Cardiovascular Disease
DBSA	Dodecylbenzenesulfonic Acid
DEA	Diethanolamine
DEAA	Diethylacrylamide
DLVO	Derjaguin, Landau, Verwey and Overbeek
DMAA	Dimethylacrylamide
DNA	Desoxyribonucleic Acid
DSC	Differential Scanning Calorimetry
DTG	Derivative Thermogravimetric
dUTP	Deoxynucleotidyl Transferase
EC	Endothelial Cell
EPS	Exopolysaccharides
e-PTFE	Expanded Poly Tetra Flourethylene
FGF	Fibrinogen Growth Factor
FTIR	Fourier-Transform Infrared
Gf	Grams-force
HAEC	Human Aortic Endothelial Cell
HCAEC	Human Coronary Artery Endothelial Cell
HUVSMCs	Human Umbilical Vein Smooth Muscle Cells
ICAM	Intercellular Adhesion Molecule

ICR	Information Collection Rule
IMS	Immunomagnetic Separation
IPN	Interpenetrating Polymer Network
IT	Information Technology
LAB	Lactic Acid Bacteria
LRR	Leucine Rich Repeats
LUMO	Lowest Unoccupied Molecular Orbital
MA	Methacrylate
MCC	Microcrystalline Cellulose
MFC	Microfibrillated Cellulose
NIR	Near Infra Red
NPs	Nanoparticles
NT	Nanotechnology
OP	Oxygen Permeability
OTR	Oxygen Transmission Rate
PCL	Polycaprolactone
PCR	Polymerase Chain Reaction
PDO	Polydioxanone
PEG	Poly (Ethylene Glycol)
PGA	Poly Glycolic Acid
PGC	Polyglecaprone
PLA	Poly Lactic Acid
PMBU	Phosphorylcholine-co-Methacryloyloxyethyl Butyl-urethane
PO	Plant Oils
POH	Prepared Hybrid Materials
PPy	Polypyrrole
PU	Polyurethane
QMRA	Quantitative Microbial Risk Assessment
SEM	Scanning Electron Microscopy
SERS	Surface-Enhanced Raman Scattering
sHPMC	Silylated Hydroxypropylmethycellulose
SIS	Small intestinal sin mucosa
SMC	Smooth Muscle Cell
SNR	Signal-to-Noise Ratio
STM	Scanning Tunneling Microscopy
TEM	Transmission Electron Microscopy
TEOS	Tetraethoxy Orthosilane

TGA	Thermogravimetric Analysis
TPF	Turmeric Petiole Fiber
US	Ultrasound
UTM	Universal Testing Machine
VEGF	Vascular Endothelial Growth Factor
WAXD	Wide Angle X-Ray Diffraction
WVP	Water Vapor Permeability

PREFACE

Natural polymers and biomaterials have always played a very important role in our lives, and the research in this field has increased tremendously over the last few decades and has led to many technological and commercial developments. ICNP 2012 – Third International Conference on Natural Polymers, Bio-Polymers, Bio-Materials, Their Composites, Blends, IPNs, Polyelectrolytes and Gels: Macro to Nano Scales took place at Mahatma Gandhi University, Kottayam, Kerala, India, on October 26, 27, and 28, 2012. It was jointly organized by Beijing University of Chemical Technology, China. The conference aimed to emphasize interdisciplinary research on processing, morphology, structure, and properties of natural polymers, biomaterials, biopolymers, their blends, composites, IPNS, and gels from macro to nano scales and their applications in medicine, automotive, civil, chemical, and aerospace, computer, and marine engineering. During the three-day conference, distinguished scientists specializing in various disciplines discussed recent advances, difficulties, and breakthroughs in the field of natural polymers and biomaterials. The conference included keynote addresses, a number of plenary sessions, invited talks and contributed lectures focusing on the diverse aspects of natural polymers and biomaterials. It included discussions on recent advances, difficulties, and breakthroughs in the field of natural polymers and biomaterials. Bio-macromolecules, bio-polymers, bio-materials, natural polymers, biocomposites, bio-nanocomposites, micro and nano blends based on natural polymers, bio-degradable polymers, interphases in composites and blends, IPNs, rheology of natural polymers, processing of natural polymers, degradation and stabilization of natural polymer systems, biomass, recycling of natural polymers were among the wide array of topics covered in the conference. Additionally, there was a poster session with more than 50 posters to encourage budding scientists and researchers in this field. The conference had over 200 delegates from all over the world.

This book, titled *Biomaterial Applications: Macro to Nano Scales*, is a collection of chapters from the delegates who presented their papers during the conference. The book chapters include a wide variety of topics in natural polymers, biomaterial, composites, and their applications.

We appreciate the efforts and enthusiasm of the contributing authors for this book, and acknowledge those who were prepared to contribute but were unable to do so at this time. We trust that this book will help in stimulating new ideas, methods, and applications in the dynamic area of research. The guest editors are Dr. Sabu Thomas, International and Inter University Centre for Nanoscience and Nanotechnology, Mahatma Gandhi University, Kottayam, India; Dr. Yang Weimin, Beijing University of Chemical Technology, China; Dr. Nandakumar Kalarikkal, International and Inter University Centre for Nanoscience and Nanotechnology, Mahatma Gandhi University, Kottayam, India; and S. Snigdha, International and Inter University Centre for Nanoscience and Nanotechnology, Mahatma Gandhi University, Kottayam, India.

We would like to thank all who kindly contributed chapters for this book and the editors of Apple Academic Press, Inc., for their kind help and cooperation. We are also indebted to the Apple Academic Press, Inc., editorial office and the publishing and production teams for their assistance in preparation and publication of this book.

—Dr. Sabu Thomas

CHAPTER 1

GREEN ORGANIC-INORGANIC HYBRID MATERIAL FROM PLANT OIL POLYOL

ERAM SHARMIN, MUDSSER AZAM, FAHMINA ZAFAR, DEEWAN AKRAM, QAZI MOHD. RIZWANUL HAQ, and SHARIF AHMAD

CONTENTS

Abstract 2
1.1   Introduction 2
1.2   Materials 3
1.3   Synthesis of ALP and POH 4
1.4   Instrumentation and Methods 4
1.5   Results and Discussion 5
1.6   Conclusion 12
Acknowledgments 12
Keywords 12
References 13

ABSTRACT

Plant oils (PO) are considered as environment friendly, domestically abundant, cost effective renewable resources. They contain functional groups (double bonds, active methylenes, hydroxyls, oxiranes), which are capable of chemical transformations, yielding host of monomers or polymers. PO polymers generally lack properties of rigidity, strength and thermal stability. In this context, over the past few decades, several modifications of PO polymers as organic-inorganic hybrid, composite and nanocomposite materials have been accomplished, which have shown proven high performance characteristics over their virgin and petro-based counterparts. In this chapter, we have presented hybrid materials (POH) from PO derived polyol and tetraethoxy orthosilane (TEOS) by microwave (MW) assisted synthesis at lower preparation time relative to their counterparts synthesized through conventional method. Some preliminary investigative studies have been carried out to assess the structure, morphology, water solubility and antibacterial behavior of the said POH. These POH are foreseen as prospective greener hybrid materials for coatings and paints.

1.1 INTRODUCTION

Recent advances in polymer science have been focused on the development of cost effective, environment friendly and high performance polymers from green resources en route Green Chemistry. The "green resources" include starch, cellulose, chitosan, plant oils [PO], and others. PO are considered as the most environment friendly, domestically abundant, cost effective bio-based precursors to polymers. They contain functional groups (double bonds, active methylenes, hydroxyls, oxiranes), which are capable of undergoing several derivatization reactions, yielding a host of monomers/polymers, for example, diols, epoxies, polyols, polyesters, polyester amides, polyurethanes. These polymers find versatile applications as adhesives, lubricants, antimicrobial agents, paints and coatings [1, 2]. PO polymers generally lack properties of rigidity, strength and thermal stability. In the past few decades, several modifications of PO polymers as organic-inorganic hybrid, composite and nanocomposite materials have appeared. These have shown improved performance characteristics over their virgin as well as petro-based counterparts [3, 4].

PO based organic-inorganic hybrids [POH] consist of nano sized inorganic reinforcements in PO derived organic matrix (e.g., alkyds, epoxies, polyols, polyester amides and polyurethanes). POH may be prepared by chemical reactions such as hydrolysis-condensation reaction between organic matrix and metal alkoxides, for example, $Zr(OR)_4$, $Al(OR)_3$, $Ti(OR)_4$, $Sn(OR)_4$, $Si(OR)_4$), by mixing preformed metal nanoparticles, for example, Ag, Cu, Au, by in situ generation of inorganic nano reinforcements or by the inclusion of layered silicates (nanoclay) [5–12] in PO derivatives. They exhibit excellent coating properties (mechanical properties, hardness, corrosion resistance, flexibility, and impact resistance) attributed to the organic and inorganic components, respectively. Literature scanning reveals that the preparation of POH generally occurs via cumbersome processes involving multistep syntheses and cure schedules at elevated temperatures with prolonged synthesis and processing times and consuming ample of solvents [6–16]. Thus, to overcome the aforementioned disadvantages, a greener, energy efficient, environment friendly approach is urgently needed, aimed at reducing multiple preparation steps and time, minimizing or eliminating the use of hazardous solvents or persuading the use of greener solvents such as water [17–25], complying with the principles of Green Chemistry.

In this chapter, we have prepared Linseed oil (LO) based POH through an energy efficient process, that is, microwave (MW) assisted synthesis technique, using polyol from LO [ALP] [26] as organic and tetraethoxyorthosilane (TEOS) as inorganic precursor via typical hydrolysis-condensation reaction. The approach presents a green protocol for the synthesis of hybrid materials, that is, (i) the use of bio-resource PO as a raw material, (ii) through MW (dielectric) heating which minimizes energy consumption and reduces preparation time [27], and (iii) using an environmentally benign solvent- water (in minimum quantity) [28], representing an excellent example of "green," "nano" hybrid material.

1.2 MATERIALS

LO (Pioneer in-organics, Delhi), hydrogen peroxide (30%), sulfuric acid, glacial acetic acid, benzene (Merck, India), sodium metal, diethyl ether, diethanolamine (DEA) (s.d. fine chemicals), (TEOS) (Merck, Germany) were used as received.

1.3   SYNTHESIS OF ALP AND POH

ALP was prepared by a two-step process consisting of hydroxylation of LO, followed by amidation [26]. In this method, firstly, a typical hydroxylation procedure was carried out with hydrogen peroxide (30%), sulfuric acid, and glacial acetic acid that produced LO polyol from virgin LO [29]. LO polyol, so obtained, produced ALP, by simple amidation process in presence of sodium methoxide and DEA [29, 30].

About 10 g of ALP, 5%, 10%, 15% of TEOS (percent by weight of ALP) (prehydrolyzed at room temperature in presence of water and HCl maintaining the pH of TEOS-water-acid mixture as 1.35) and 1 mL distilled water were taken in an Erlenmeyer flask and placed in the MW oven for two minutes maintaining the temperature at 50 °C. The reaction mixture was observed for clarity, perceptible changes such as abnormal rise in viscosity or phase separation, if any, throughout the course of the reaction. POH were obtained as clear, free flowing and yellowish brown colored liquids. POH were named as 5-POH, 10-POH and 15-POH, where the numbers indicate the percent loading of TEOS.

1.4   INSTRUMENTATION AND METHODS

The synthesis was carried out in LG microwave oven (model LG MS 1927C, MW frequency – 2500 MHz; power source –230 V ~50 Hz, energy output – 800 W; input power – 1200 W).

FTIR spectra were taken on IR Affinity–1 CE (Shimadzu corporation analytical and measuring instrument division, Kyoto, Japan) by placing the samples in between two zinc selenide windows. Spectra were taken from an average of 40 scans for each sample.

Morphological studies were carried out by transmission electron microscopy (TEM) (Morgagni 268-D TEM FEI instrument, Netherlands) on a carbon coated copper grid. The test sample was diluted well in ethanol and submerged in an ultrasonic bath for sonification for 30 min. Next, a drop of this solution was placed on a Carbon Type-B (carbon film supported) 200 mesh copper grid by a micro pipette. The grid was allowed to dry well before being used for TEM analysis.

Solubility in water was assessed by dissolving the samples in water (percent by weight) in standard sample bottles at room temperature (28 °C–30 °C) monitored through timely vigilant observation.

Transparency characteristic was visually observed by simply applying these POH on glass slides.

Preliminary antibacterial studies of POH were performed by disk diffusion assay against *Staphylococcus aureus* (MTCC 902) and *Pseudomonas aeruginosa* (MTCC 2453).

1.5   RESULTS AND DISCUSSION

Hydroxylation of LO was carried out in situ that produced plain LO polyol by the introduction of (Trans) hydroxyl groups on LO backbone. Plain LO polyol further by base (sodium methoxide) catalyzed amidation reaction with DEA, resulted in the formation of ALP and glycerol. The advantage of this step was the introduction of amide groups along with additional hydroxyl functionalities in the polyol backbone.

The synthesis of POH was accomplished in two steps in the presence of water and glycerol (obtained during the preparation of ALP), as solvents [26]. The first step involved hydrolysis of TEOS wherein silanols were generated that condensed with hydroxyl groups of ALP and glycerol in the second step, resulting in the anchorage of –O–Si–O– in LO ALP forming POH (Fig. 1.1) [5, 31]. Here, 5-POH and 10-POH were obtained as clear, glossy liquids while 15-POH was relatively foggy attributed to higher content of TEOS. The said POH prepared through MW technique were obtained in 2 min, while their conventionally prepared counterparts were produced in 2 h at 50 °C. In MW assisted reactions (dielectric heating), due to rotation, friction and collision of (polar) molecules, under applied electromagnetic field, excessive heat is produced; the chemical reaction occurs in much reduced time.

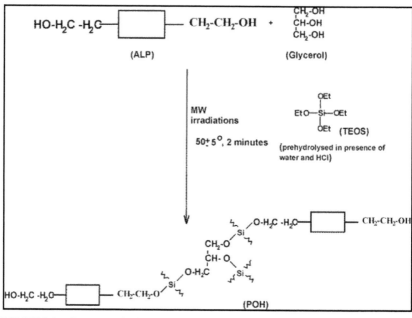

FIGURE 1.1   Structure of POH.

1.5.1   FTIR

FTIR spectra of 5-POH, 10-POH and 15-POH (Fig. 1.2a–c) show absorption bands characteristic for –OH appearing at 3375–3360 cm^{-1} with slight depression as the content of TEOS increased, correlated to the chemical reaction (hydrolysis–condensation) between –OH of polyol and TEOS [31, 32]. Relatively intense bands are observed at 862 cm^{-1}, 1066 cm^{-1} and 1120 cm^{-1}. These absorption bands gradually appear more intense with increased TEOS content, supporting the formation of –Si–O–Si– network in polyol [32], which are engulfed by the remnant polyol backbone. These absorption bands support the incorporation of –O–Si–O– in ALP backbone forming POH [5, 26, 31–33].

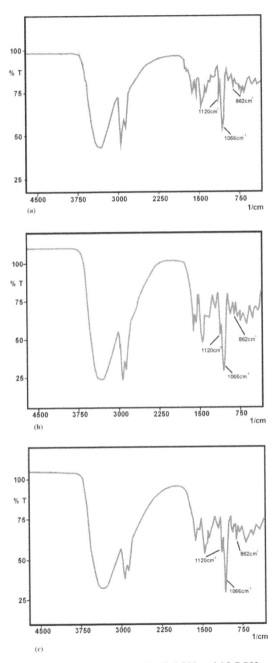

FIGURE 1.2   FTIR spectra of (a) 5-POH, (b) 10-POH and 15-POH.

1.5.2   TEM

TEM micrographs of POH are given in Fig. 1.3a–c. The visible gray region is the ALP matrix while the scattered dark points or globules observed are the silica nanoparticles. The nanoparticles reveal size ranging from 4 nm–24 nm, with the highest distribution of particles of dimensions 14 nm–16 nm. These exist as spherical particles with distinct boundaries and seem to be embedded within the matrices or engulfed by aliphatic fatty chains of parent PO. These occur as unagglomerated particles since a portion of polyol backbone is chemically bonded to nanosilica particles (also supported by FTIR analysis) preventing their agglomeration [34–38]. The study confirms the formation of organic-inorganic nanohybrids from LO.

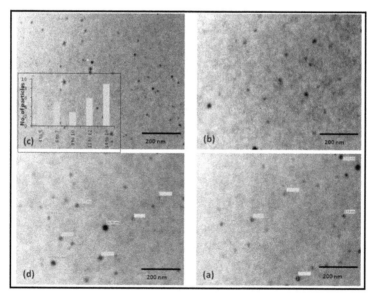

FIGURE 1.3   TEM micrographs of (a) 5-POH and (b), (c), (d) 10-POH.

1.5.3   SOLUBILITY IN WATER

Freshly prepared 5-POH and 10-POH solutions (100%–20%) in water (Fig. 1.4a) were obtained as clear solutions at room temperature (28 °C–30 °C). These remained as clear homogenous solutions even after 12 h of vigilant observations at the given temperature.

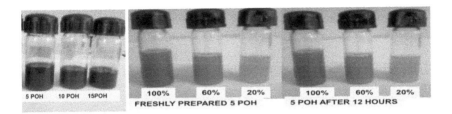

5 POH   10 POH   15POH

100%   60%   20%   100%   60%   20%
FRESHLY PREPARED 5 POH   5 POH AFTER 12 HOURS

(a)

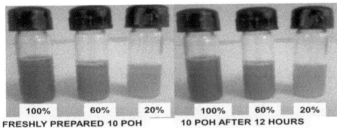

100%   60%   20%   100%   60%   20%
FRESHLY PREPARED 10 POH   10 POH AFTER 12 HOURS

(b)

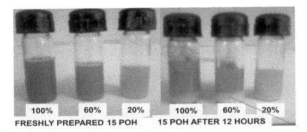

100%   60%   20%   100%   60%   20%
FRESHLY PREPARED 15 POH   15 POH AFTER 12 HOURS   whitening appears

(c)

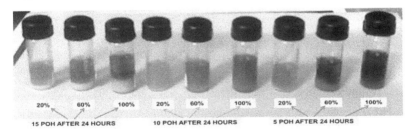

20%   60%   100%   20%   60%   100%   20%   60%   100%
15 POH AFTER 24 HOURS   10 POH AFTER 24 HOURS   5 POH AFTER 24 HOURS

(d)

FIGURE 1.4   Solubility of 5-POH, 10-POH and 15-POH in water.

1.5.4   TRANSPARENCY CHARACTERISTIC

POH were applied on glass slides and placed on a plain white sheet with letters marked as "N," "A" and "R." These letters were clearly visible through POH coated glass slides. Thus, POH produced fully transparent films as evident in Fig. 1.5.

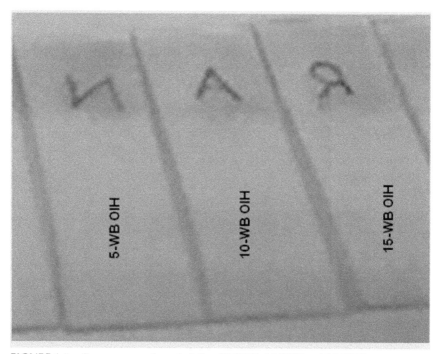

FIGURE 1.5   Transparency characteristic of 5-POH, 10-POH and 15-POH.

1.5.5   ANTIBACTERIAL BEHAVIOR

Determination of antibacterial sensitivity pattern of POH was performed by Kirby Bauer method (i.e., disk diffusion assay) against *Staphylococcus aureus* (MTCC 902) and *Pseudomonas aeruginosa* (MTCC 2453) [40, 41]. Mueller-Hinton agar plates were inoculated by the test organisms for a uniform loan of culture. Discs containing 40 mg of the test sample were placed at appropriate distance from the center. Para-film sealed plates were incubated for 12–14 h at 37 °C. After incubation, zone of inhibition was measured and found to be 10 mm and 9 mm for *Staphylococcus aureus* and *Pseudomonas aeruginosa, respectively.* Sensitivity pattern of same isolates were tested for wide range of antibiotics having different mode of action (Table 1.1).

TABLE 1.1   Antibiogram of Bacterial Isolates for Tested Antibiotics

Antibacterial agent	MTCC 902	MTCC 2453
POH	10	9
Ampicillin/sulbactum	41 (S)	6 (R)
Levofloxacin	28 (S)	28 (S)
Rifampicin	26 (S)	6 (R)
Tetracycline	26 (S)	15 (S)
Polymixin B	8 (R)	15 (S)
Trimethoprim	20 (S)	6 (R)
Ofloxacin	21 (S)	21 (S)
Aztreonam	24 (S)	30 (S)
Tobramycin	23 (S)	35 (S)
Ciprofloxacin	25 (S)	36 (S)
Cefozolin	32 (S)	6 (R)
Ertapenem	29 (S)	14 (R)
Ceftazidime	16 (R)	20 (R)
Cefotaxime	33 (S)	24 (R)

Zone of Inhibition: diameter in mm   S–Susceptible; R–Resistant.

These results showed that POH was efficient enough to inhibit the growth of the given microbes, resistant to a number of antibiotics tested for their sensitivity pattern (Table 1.1). Our preliminary investigations on antibacterial behavior of POH by disc diffusion method revealed that this bactericidal activity may be attributed to the structural build-up of POH. Further studies on the mode of antibacterial action of POH are still needed to be carried out.

1.6 CONCLUSION

LO based nanostructured POH were prepared by MW assisted preparation method from amidated Linseed polyol and TEOS. The approach dealt with the use of a bio-resource, simple, safe, less time consuming and energy efficient MW assisted preparation method, without the use of any hazardous solvents but water, with no complex chemical transformations or side reactions complying with the principles of Green Chemistry. Our studies revealed that the said LO based POH could be used as water soluble materials for time period not exceeding 24 h at ambient temperature (28 °C–30 °C). Thus, what still remains on tenterhooks and unanswered is the 100% solubility and 100% stability (as water soluble materials) of POH for largely extended time periods at room temperature.

ACKNOWLEDGMENTS

Dr. Eram Sharmin (Pool Officer) and Dr. Fahmina Zafar (Pool Officer) acknowledge the Council of Scientific and Industrial Research (New Delhi, India) for Senior Research Associate ships against Grant Nos. 13(8464–A)/2011 Pool and 13(8385-A)/Pool/2010, respectively. Dr. Fahmina Zafar is also thankful to University Grant Commission, India for Dr. D. S. Kothari Post Doctoral Fellowship, Ref. # F.4/2006(BSR)/13-986/2013(BSR). Dr. Eram Sharmin and Dr. Fahmina Zafar are thankful to the Head, Department of Chemistry (Jamia Millia Islamia, New Delhi) for providing facilities to carry out the research work. Mudsser Azam is thankful to the Council of Scientific and Industrial Research (New Delhi, India) for financial assistance against Grant No. 09/466(0136)/2011-EMR-I

KEYWORDS

- **Organic-Inorganic Hybrids**
- **Polyol**
- **Vegetable Oils**

REFERENCES

1. Espinosa, L. M., & Meier, M. A. R. (2011). Eur. Polym. J. 47, 837.
2. Lligadas, G., Ronda, J. C., Galia, M., & Ca´ diz, V. (2010). Biomacromol., 11, 2825.
3. Tsujimoto, T., Uyama, H., & Kobayashi, S. (2010). Polym Degrad. Stab, 95, 1399.
4. Bordes, P., Pollet, E., & Avérous, L. (2009). Prog Polym Sci., 34, 125.
5. Akram, D., Ahmad, S., Sharmin, E., & Ahmad, S. (2010). Macromol Chem. Phys, 211, 412.
6. Chattopadhyay, D. K., & Raju, K. V. S. N. (2007). Prog. Polym Sci., 32, 352.
7. Kumar, A., Vemula, P. K., Ajayan, P. M., & John, G. (2008). Nat Mater, 7, 236.
8. Konwar, U., Karak, N., & Mandal, M. (2010). Prog Org. Coat., 68, 265.
9. De Luca, M. A., Martinelli, M., & Barbieri, C. C. T. (2009). Prog Org. Coat., 65, 375.
10. Dutta, S., Karak, N., Saikia, J. P., & Konwar, B. K. (2009). Bioresource Technol., 100, 6391.
11. Deka, H., & Karak, N. (2009). Nanoscale Res Lett. 4, 758.
12. Uyama, H., Kuwabara, M., Tsujimoto, T., Nakano, M., Usuki, A., & Kobayashi, S., (2004). Macromol Biosci. 4. 354.
13. Becchi, D. M., De Luca, M. A., Martinelli, M., & Mitidieri, S. (2011). J. Am Oil Chem. Soc., 88, 101.
14. Martinelli, M., De Luca, M. A., Becchi, D. M., & Mitidieri S. (2009). J. Sol-Gel Sci., Technol., 52, 202.
15. De Luca, M. A., Martinelli, M., Jacobi, M. M., Becker, P. L., & Ferrão, M. F. (2006). J. Am Oil Chem. Soc., 83, 147.
16. Brasil, M. C., Gerbase, A., De Luca, M. A., Grego´rio, J. R. (2007). J. Am Oil Chem. Soc., 84: 289.
17. Athawale, V. D., & Nimbalkar, R. V. (2010). J. Am Oil Chem. Soc. 88, 159.
18. Balakrishna, R. S., & Sivasamban, M. A. (1971). J. Colour Soc., 10, 2.
19. Aigbodion, A. I., Okieimen, F. E., Obazee, E. O., & Bakare, I. O. (2003). Prog Org. Coat. 46, 28.
20. Baolian, Ni. E., Liting Yang, E., Chengshuang Wang, E., Linyun Wang, E., David, E., Finlow, (2010). J. Therm Anal Calorim 100, 239.
21. Gao, C., Xu, X., Ni, J., Lin, W., Zheng, Q. (2009). Polym Eng and Sci, 49, 162.
22. Aigbodion, A. I., Okiemien, F. E., Ikhuoria, E. U., Bakare, I. O., & Obazee, E. O. (2003). J. Appl. Polym. Sci 89, 3256.
23. Lu, Y. & Larock, R. C. (**2007**). *Biomacromol, 8*, 3108.
24. Hu, Y. S., Tao, Y., & Hu, C. P. (**2001**). *Biomacromol.* 2, 80.
25. Lu, Y. & Larock, R. C. (2010). Prog.Org.Coat, 69, 37.
26. Eram Sharmin, Akram, D., Zafar, F., & Ahmad, S. (2009) "Polyol from Linseed Oil for Waterborne Coatings, Synthesis and Characterization," Presented at International conference "Polymer Science & Technology, Vision & Scenario" (APA–2009) at New Delhi, India on Dec. 17–20.
27. Sarma, R., & Prajapati, D. (2010). Green Chem., 13, 718.
28. Andrade, C. K. Z., & Alves, L. M. (2005). Curr. Org. Chem., 9, 195.
29. Sharmin, E. Ashraf, S. M., & Ahmad, S. (2007). Int.J. Biol. Macromol, 40, 407.

30. Kashif, M., Sharmin, E., Zafar, F., & Ahmad, S. (1989) J. Am. Oil Chem. Soc. 88, (2011).
31. Fujiwara, M., Kojima, K., Tanaka, Y., & Nomura, R. (2004). J. Mater Chem, 14, 1195.
32. Lee, T. M., & Mia, C.C.M. (2006). J. Polym. Sci. Part A: Polym Chem., 44, 757.
33. Wang, L., Tian, Y., Ding, H., & Li, J. (2006). Eur Polym J. 42, 2921.
34. Chen, Y., Zhou, S., Yang, H., Ge, G., & Wu, L. (2004). J. Colloid Int. Sci. 279, 370.
35. Jena, K. K., & Raju, K. V. S. N. (2008). Ind Eng. Chem. Res., 47, 9214.
36. Zhai, L., Liu, R., Peng, F., Zhang, Y., Zhong, K., Yuan, J., Lan, Y. (2013). J. Appl. Polym. Sci., 128, 1715.
37. Lligadas, G., Callau, L., Ronda, J. C., Galia, M., & Cadiz, V. (2005). J. Polym. Sci., Part A: Polym. Chem., 43, 6295.
38. Asif, A., & Shi, W. (2003). Eur Polym J., 39, 933.
39. Xia, Y., & Larock, R. C. (2011). Macromol Rapid Commun., 32, 1331.
40. Antimicrobial Susceptibility Testing by Donald C. Sockett DVM, PhD, Wisconsin Veterinary Diagnostic Laboratory, 01/04/13.
41. Clinical and Laboratory Standard Institute (2006) Performance standards for antimicrobial susceptibility testing. In: 16th informational supplement, Wayne, PA.

CHAPTER 2

BIO-HYBRID 3D TUBULAR SCAFFOLDS FOR VASCULAR TISSUE ENGINEERING—A MATERIALS PERSPECTIVE

HARSH PATEL, ROMAN GARCIA, and VINOY THOMAS

CONTENTS

2.1   Introduction... 16

2.2   Bio-Hybrid Vascular Graft Based on Collagen/Synthetic Polymers 19

2.3   Bio-Hybrid Vascular Grafts Based on Elastin 23

2.4   Bio-Hybrid Vascular Graft Based on Collagen/Elastin/ Synthetic Polymers 25

2.5   Bio-Hybrid Vascular Grafts Based on Gelatin/Synthetic Polymers 26

2.6   Bio-Hybrid Vascular Grafts Based on Fibrin/Fibrinogen 30

2.7   Bio-Hybrid Vascular Grafts Based on Bacterial Cellulose........... 32

2.8   Bio-Hybrid Vascular Graft Based on Silk Fibroin/Synthetic Polymers 34

2.9   Multi-Layered Bio-Hybrid Vascular Grafts 37

2.10 Summary... 38

2.11 Acknowledgments 40

Keywords 41

References... 41

2.1   INTRODUCTION

Cardiovascular disease (CVD) remains one of the leading causes of death around the world [1]. This global health concern has prompted advances in surgical and medical technologies. However, an increasing percentage of the population continues to be at risk and eventually fall victim to this debilitating disease. While the most common cause of blood vessel failure in this case, is atherosclerosis, whereby plaque buildup within the lumen results in hindered blood flow, injury, disease, or inflammation can also result in a compromised vessel [2]. Routinely, a comparable native vessel (ex. internal thoracic arteries, radial arteries, and saphenous veins) will be excised and relocated to replace the occluded vessel in a procedure known as a coronary artery bypass graft (CABG) surgery. These autologous grafts have limited availability and depend on donor site morbidity [3]. For example, with close to a half a million coronary artery bypass operations performed annually in the US alone, more than 20% of the patients do not possess healthy graft veins for CABG [4]. Addressing this concern, researchers have increasingly looked towards scaffold-based vascular tissue engineering to produce readily available vascular grafts [5]. While synthetic polymers such as polyester (Dacron) or expanded poly tetra flourethylene (e-PTFE) have been used successfully to fabricate large diameter (>6 mm) blood vessels [6], these materials have resulted in loss of vessel patency due to thrombosis in small diameter (<6 mm) applications [7]. Furthermore, as these synthetic vascular grafts will ultimately be implanted into the patient, the biocompatibility of component materials is an important factor for the future success of this technology. To this end, vascular tissue engineers have shown a general movement toward small diameter (<6 mm) scaffolding comprised of biopolymers, those that naturally exist within vascular tissue matrix as well as naturally occurring biomaterials.

The arterial wall structure of native vessels consists of three distinct layers, intima, media, and adventitia, each having its own specific cell and protein composition. The innermost layer (intima) surrounding the vessel lumen contains a monolayer of endothelial cells (ECs), collagen type IV, and elastin. The cellular physiology of the intima aids in the prevention of thrombosis, infection, and inflammation of the blood vessel and any underlying tissue [2]. The middle layer (media) is contains collagen type I, III, elastin, proteoglycans, and smooth muscle cells (SMCs). This layer provides the unique mechanical properties necessary for the vessel to withstand in vivo pressures and forces. The outermost layer (adventitia) is primarily comprised of collagen type I along with fibroblasts. As the outermost layer, the adventitia's primary function is to provide stability to the vessel by anchoring it to nearby tissue as well as to connect with body systems via vascular and nerve networks. Each distinct matrix/cell layer works in tandem to withstand the physiologically rigorous pressure conditions of the body and maintain homeostasis. Table 2.1 lists the mechanical properties for common native vessels used for synthetic vascular graft comparison.

TABLE 2.1 Mechanical Properties for Some Native Vessels

Native vessel	Tensile Strength (MPa)	Tensile Modulus (MPa)	Strain (%)	Reference
Femoral artery	1–2	9–12	63–76	[8]
Fresh carotid artery	1.76–2.64	-	110–200	[9]
Coronary artery	1.4–11.14	-	45–99	[10]

The three dimensional tubular structure of vascular scaffolds must provide a platform for cell migration, proliferation, and differentiation while mimicking the structural and mechanical integrity of native vascular ECM. The ideal scaffold should be biocompatible, biodegradable, bioresorbable and a cellular-growth enhancing matrix. As new tissue formation starts, the scaffold should gradually degrade leaving enough space for tissue in-growth. Further, material degradation rates must be considered as any premature hydrolysis of these polymers prior to complete cellular proliferation will compromise the mechanical integrity of the scaffold. Vascular tissue scaffolds can be divided into three types based on the nature and origin: natural, synthetic, and bio-hybrid. Examples of natural scaffolds would be those made of collagen, elastin, fibronectin etc., materials which are structural and/or specialized proteins of native vascular ECM. Synthetic scaffolds can be made of different biodegradable polyester polymers such as polycaprolactone (PCL), poly lactic acid (PLA), poly glycolic acid (PGA), copolymer form of a poly D, L lactic-coglycolic acid (PLGA) etc. Bio-hybrid scaffolds use a combination of natural and synthetic polymers to achieve adequate elastic mechanical properties attainable through synthetic polymers while increasing biocompatibility and cellular recognition through the use of natural materials.

Various methods have been used to create tubular scaffolds for vascular tissue engineering. For example, researchers have engineered a completely biological vascular graft by incorporating collagen type I into a gel scaffold through the use of molds prepared with the help of silicone sleeves and hollow glass tubes [32]. Further, a procine derived small intestinal sin mucosa (SIS) has been used to generate a biological vascular scaffold or xenograft, since it is a decellularized tissue or ECM [33]. However, concerns of disease transmission and negative immune response continue to arise for vascular graft made from SIS. Another approach used to create a completely biological tissue engineered blood vessel was developed by the Augur group [34]. They made single layer cell sheets comprised of human vascular SMCs wrapped around a tubular support. Afterward, other cell sheets comprised of human fibroblasts were wrapped around of the previous SMC sheet. After a few days of cell growth, the tubular support was removed leaving a tubular-shaped biological tissue engineered vascular graft. In all these methods, the mechanical integrity of the graft has not been adequate. Furthermore, these methods produce scaffolds with low porosity limiting cell-growth and proliferation.

While many methods do exist to produce vascular scaffolds, electro spinning, a nanotechnology enabling technique, has attracted much attention due to its simplicity to fabricate tubular 3D conduits and replicate native ECM-like formation of random oriented nanofibers with large surface area to volume ratio for enhanced cell adhesion and proliferation [35]. Electro spinning is a facile technique for producing ultra-thin polymer micronano fibers using an electrostatic force [36]. Detailed reviews on the electro spinning process and the applications of electro spun biomimetic scaffolds can be seen elsewhere [37]. Electrospun scaffolds are highly porous with interconnecting pore-networks (75–90%), providing a large surface area with surface functional groups for protein and cell attachment. Electrospinning of any polymer requires optimization of two main parameters: System parameters and process parameters. System parameters consists of molecular weight, molecular weight distribution, polymer solution viscosity, and conductivity, whereas process parameters consist of electric potential, flow rate, distance between the capillary and collector, temperature, and humidity [36]. These parameters can be varied to achieve bead-free quality of fibrous mats for tissue engineering applications.

This process allows for the formation of nanofibers composed of polymer blends, natural and synthetic, all done in a rather straightforward, cost effective setup [17]. Therefore, an increased emphasis is given to review electrospun bio-hybrid scaffolds, their mechanical integrity, biocompatibility, and native cell adhesion and proliferation in this chapter.

2.2 BIO-HYBRID VASCULAR GRAFT BASED ON COLLAGEN/ SYNTHETIC POLYMERS

Since the introduction of biopolymers in vascular tissue engineering, collagen has been a favored material due its prevalence in native vascular make-up. It is made up of groups of fibrous proteins and naturally produced by mammalian cells. It is also one of the most plentiful proteins found in human body. There are almost 28 different types of collagen in the human body [38]. Collagen is a preferable biopolymer because it's low antigenicity, low inflammatory and cytotoxic response, and desirable biological and hematological properties [39]. Collagen has been used in various biomedical applications due to its biocompatibility and biodegradability [40]. As mentioned previously, it is a main component in the blood vessel matrix-structure. Therefore, researchers have strived to compose vascular grafts using collagen and gelatin (denatured collagen) with or without other biodegradable polymers. The basic structure of collagen is a three-parallel polypeptide chains formed by glycine, proline and hydroxyl proline repeating units arranged in a triple helix structure [41]. The mechanical integrity of collagen type I can be different than collagen type III or others and different tissue types have different types of collagen. In addition, collagen origin can also play a role in the final characteristics of the collagenous tissue-scaffolds derived from animal or cadaver tissues [11].

Being a vascular matrix protein, researchers started to fabricate hybrid grafts made of collagens and biodegradable synthetic polymers to impart bioactivity and to prevent scar-tissue fibrosis. Weinberg and Bell have incorporated collagen with Dacron (nonbiodegradable) to create a model for an artificial blood vessel in vitro [42]. Thereafter, the biodegradable grafts were introduced since they would degrade in vitro and in vivo as the native tissue formation starts. Widely used aliphatic polyesters in tissue engineering applications take several months to a few years to degrade. The biodegradation and cell-interaction characteristics of them can be tuned with the incorporation of collagen. For example, when collagen fibers were incorporated onto electrospun PLA fibers to mimic the properties of blood vessel [43], promising SMCs growth and mechanical integrity improvements were observed. To mimic the smooth muscle cells arrangement in the arterial wall, a preliminary graft scaffold based on PLLA was created by electrospinning method on 4 mm diameter mandrel with a single helical wind of collagen fibers embedded in biodegradable matrix [43]. After 10 days culture with human aortic SMCs, the vascular graft exhibited extensive SMCs proliferation and the orientation of the SMCs.

Because of rapid degradation and low shear-stress resistance of pure collagen scaffolds, there is a need for reliable and biocompatible synthetic polymer for preventing premature degradation of scaffolds. The longer degradation time is a useful characteristic when combined with collagen in a vascular tissue graft, as cells need longer time to proliferate and to generate ECM to mimic native tissue structure. PCL is an aliphatic polyester with a low glass transition temperature of 60 °C and low melting point 60 °C [39] and slower in degradation (half-life is ~2 years). Atala's group has combined PCL with collagen type I and seeded with bovine ECs (bECs) and bovine SMCs (bSMCs) [12]. To fabricate the scaffold, PCL and collagen type I (1:1 ratio) were dissolved in the same solvent and then the solution was electrospun into tubular conduit. They noticed formation of the confluent layer of bECs on the lumen and bSMCs on the outer surface of the scaffold, which exactly mimic the native artery structure. The mechanical properties were shown to be comparable to native vessels (listed in Table 2.2). An important mechanical characteristic which is useful in understanding whether or not the graft will survive in the arterial environment for vascular graft is its burst pressure. The gold standard for CABG is a saphenous vein, which has a burst pressure of 1680 ± 370 mmHg [44]. The burst pressure of the composite PCL/collagen (1:1) scaffold was 4912 ± 155 mmHg with an acceptable suture retention strength (above 2.0 N) [45]. Figure 2.1 details the vascular graft with uniform wall-thickness made of PCL/collagen fibers with electrospinning technique [12]. It also illustrates the surface morphology of the scaffold exhibiting randomly oriented fibers and interconnected pores throughout the thickness.

TABLE 2.2 Mechanical Properties of Different Bio-Hybrid Compositions Used in Vascular Grafts

Scaffold Composition	Diameter of the fibers	Tensile Strength (MPa)	Tensile Modulus (MPa)	Strain at break (%)	Reference
Collagen type I	100–730 nm	1.5 ± 0.2	52.3 ± 5.2	–	[11]
PCL/collagen (wet)	520 nm	4.0±0.4	2.7 ± 1.2	140 ± 13	[12]
PCL/collagen I/ collagen III	–	18	7.79	–	[13]
P(LLA- CL)/collagen I	470 ± 130 nm	6.27 ± 1.38	43.99 ± 4.04	176 ± 49	[14]
PDO/collagen I		4.6–6.7	7.6–18.0	56.5–186.4	[15]
Gelatin	460 ± 148–517 ± 241 nm	8–12	426 ± 39	80–100	[16]
Gelatin/PCL	–	1.29	30.8	1.38	[17]
PLCL/gelatin (50:50)	200 nm – 1.3 µm	5.1 ± 0.8	340 ±8.5	86 ± 19	[18]
PLGA/gelatin/ elastin	317 ± 46 nm	130 ± 7	770 ± 131	–	[19]
PGA/gelatin (30%)	863.96 ± 265.09 nm	–	32	32	[20]
Polyglyconate/ gelatin/elastin	200–400 nm	2.71 ± 0.2	20.4 ± 3	140 ± 3	[21]
Polydioxanone/ gelatin/elastin	300–500 nm	1.77 ± 0.2	5.74 ± 3	75.08 ±10	[22]
Fibrinogen	80–700 nm	2	80	–	[23]
Fibrinogen/ PLLA-CL (20:80)	305 ± 78 nm	10.1 ± 0.4	–	118.8 ± 12.3	[24]
PLGA/fibrin	50 nm - 4 µm	0.651 ± 0.24	–	105 ± 10	[25]
Bacterial Cellulose		311 ± 29	–	–	[26]
B. mori Silk	9.3 ± 0.3 µm	740	10000	20	[27]
B. mori silk/ PEO	–	2.42±0.48	2.45±0.47	–	[28]
Silk fibroin/gelatin	143 ± 36 nm	1.12 ± 0.11	–	30.55 ± 3.46	[29]
PLA/SF – gelatin	–	1.28 ± 0.21	–	41.11 ± 2.17	[29]
Collagen/chitosan/ PLLA-CL (25:5:75)	409 ± 120 nm	16.9 ± 2.9	10.3 ± 1.1	112 ± 11	[30]
PEUU/PMBU (15%)	525±162 nm	7 ± 1	7 ± 3	342 ± 43	[31]

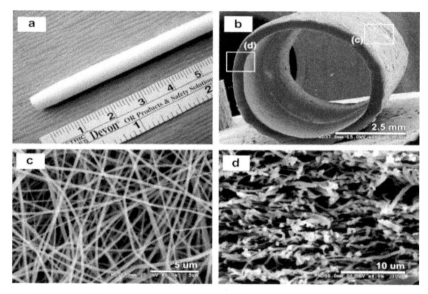

FIGURE 2.1   (a) The gross appearance and SEM images of electrospun PCL/collagen composite scaffolds: (b) entire (x18), (c) surface (x6.0K), and (d) cross-sectional (x4.0K) morphologies. Reprinted from Biomaterials, Volume 29/Issue 19, Sang Jin Lee, Jie Liu, Se Heang Oh, Shay Soker, Anthony Atala, James J. Yoo, Development of a composite vascular scaffolding system that withstands physiological vascular conditions, 2891–2898, 2008, with permission from Elsevier.

In another study, collagen was combined with the poly (L-lactide) co-poly (caprolactone) (PLLA-CL) and electrospun into nanofibers to create a mesh like structure to mimic native ECM both morphologically and bio-chemically [46]. In contrast with the previous method, the PLLA CL fibers were electrospun first and then treated with air plasma treatment for 5 min. Plasma treatment has been used to successfully increased hydrophilicity of the material, which aids in cell attachment [47]. The scaffold was subsequently coated with a collagen solution. The scaffold exhibited porosity between 60–70%. Fiber diameter, tensile strength, and tensile modulus (listed in Table 2.2) were promising. Human coronary artery endothelial cells (HCAECs) were seeded on these nanofibers to promote endothelialization, and it provided promising results for potential use for vascular tissue engineering. Further regeneration studies using bioreactor (HAECs seeded inside the lumen of the tube) and implantation in rabbits, the tubular scaffolds coated by the collagen type 1 [48] exhibited promising results that scaffolds were maintained structure integrity up to 7 weeks in vivo. Figure 2.2 illustrates the human aortic endothelial cells (HAECs) spreading and attachment on the large surface area of the electrospun scaffold made of PCL/collagen [48].

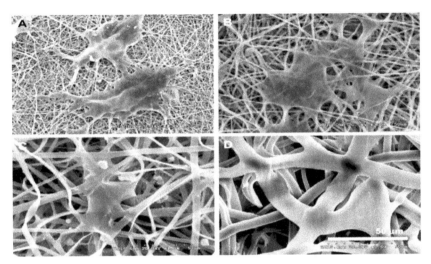

FIGURE 2.2 (A) FC1, 0.27 ± 0.09, (B) FC2, 1.00 ± 0.15, (C) FC3, 2.39 ± 0.69, and (D) FC4, 4.45 ± 0.81 mm. Scale bar indicates 50 mm (magnification; x1K). EC morphologies on the scaffolds with various fiber diameters were observed. Cell adhesion and spreading can be guided by the direction of the smaller fibers. Reprinted from Biomaterials, Volume 31, Issue 15, Young Min Ju, Jin San Choi, Anthony Atala, James J. Yoo, Sang Jin Lee, Bilayered scaffold for engineering cellularized blood vessels, 4313–4321,2010, with the permission from Elsevier. SEM images of HAEC adhesion and spreading onto electrospun PCL/collagen fibrous scaffolds with four different fiber diameters at 3 days after cell seeding.

2.3 BIO-HYBRID VASCULAR GRAFTS BASED ON ELASTIN

Elastin is a native protein widely used in electrospun vascular tissue scaffolding as it is a critical component in the complex architecture of native vascular extracellular matrix. The protein is found within the intima and media layers of the blood vessel wall. Elastin fibers can maintain their structure even after extension of approximately 140% [49]. It plays a critical role in preserving the shape and elasticity of tissues under stress, therefore integral in maintaining the biomechanical properties of native vascular tissue [50]. Scaffolds were made by using pure elastin [51]. While pure elastin has been electrospun to achieve fiber diameters within range of native matrices [50], scaffolds comprised of pure elastin alone do not exhibit adequate properties of a successful graft. Gelation and hydrolysis rapidly degrade proteins such as elastin when placed in culture media [52]. Therefore, elastin-based constructs alone do not withstand the mechanical stress common to vascular tissue [52]. Alternatively, elastin is used as a component in a blend of polymers, allowing elastin's unique properties to be expressed in relation to other materials. For example, an elastin-collagen blend in varying concentrations was electrospun in order to fabricate a tubular scaffold with layered structure [50]. A solution of 80:20 collagen type I: elastin was electrospun around a 4 mm tubular mandrel, which formed the outermost adventitia layer. At this ratio and at a concentration of 0.083 g/ml an average fiber diameter of 0.49 ± 0.22 μm was observed. A subsequent layer of 30:70 collagen type I: elastin was formed around a 2 mm tubular scaffold. The insertion of this smaller tube into the 4 mm tubular scaffold formed the vascular graft [50].

Expanding on Boland's preliminary study of electrospun elastin and collagen, Li et al. compared solubilized alpha-elastin and recombinant human tropoelastin [16]. Increasing the concentration of elastin and tropoelastin from 10% to 20% in spinning solution has yielded continuous, uniform fibers free of beads. Both biopolymers produced fibers of a greater size than gelatin and collagen. Tropoelastin fibers exhibited widths 2–3 times greater than those of elastin under similar conditions. Elastin and tropoelastin fibers under SEM imaging appeared to be flat and ribbon-shaped which was observed to be independent of variant electrospinning parameters. The native wave-like pattern of elastin remained after electrospinning. Tensile testing showed a tensile strength of 1.6 MPa for elastin and 13 MPa for tropoelastin. Li et al. concluded that tropoelastin would be a more advantageous biopolymer than soluble alpha elastin when engineering vascular grafts. Cell-scaffold interactions were favorable in elastin and tropoelastin scaffolds. Another study of electrospinning of elastin, NaCl, and PEO (poly ethylene oxide) blend comprised of elastin 5% (w/v) with a 5:1 ratio of elastin to PEO at a voltage of 10 kV and infusion rate of 50 μL/min, a fibrous matt with an average fiber diameter was 500 nm was obtained [53]. Substantial splaying of elastin/PEO fibers had contributed to the small fiber diameters observed in this experiment. Increasing the concentration of elastin has decreased the surface tension of the jet during electrospinning, which leads to decreased bead formation. However, an increase in fiber diameter was observed with increasing elastin concentration [53].

2.4 BIO-HYBRID VASCULAR GRAFT BASED ON COLLAGEN/ ELASTIN/SYNTHETIC POLYMERS

Collagen has been combined with the elastin to recapitulate the biomechanical and biochemical properties of native ECM. An artery's mechanical stability greatly depends on the amount of collagen present; however, the elasticity and the ability to recover from the deformation are dependent on the presence of elastin. Therefore, a hybrid type of scaffold was prepared by electrospinning method but, incorporating polydioxanone (PDO), elastin, and collagen [54] with different compositions 50:25:25 and 45:45:10 (PDO:elastin:collagen). Buttafoco et al. have spun collagen type I (from calf skin) and elastin (from bovine neck ligament mixed) together with the high molecular weight of poly (ethylene oxide) PEO to electrospun nanofibers (ranging from 220 to 600 nm diameter) scaffolds [53]. After electrospinning, the meshes were stabilized by crosslinking using a mixture of N-(3-dimethylaminopropyl)-N''-EDC and N-hydroxysuccinimide (N-NHS) [55]. Interestingly, another arterial vascular graft composed by multi lamellar structure from synthetic collagen microfibers and a recombinant elastin like protein [56] in which the synthetic collagen fibers settled into parallel arrays embedded within the elastin, and then the entire matrix is rolled into tubular form to create a vascular graft. The mechanical characteristics such as burst pressure and suture retention were measured to be 239–2760 mm Hg and 35–192 grams-force (gf), respectively. In short, postfabrication methods such as cross-linking affect the mechanical strength due to fiber architecture, fiber spacing and orientation.

Electrospun nanofibers scaffolds made with Poly (D, L-lactide-coglycolide, 50:50 ratio) PLGA and incorporated collagen type I from calf skin and elastin were examined for vascular graft application [57]. Figure 2.3 shows the scaffold made of PLGA, collagen and elastin with different ratios and the SEM image of their nanofibers [57]. Important to note, when subjected to a burst pressure 12 times higher than normal systolic blood pressure, the scaffold did not rupture [57].

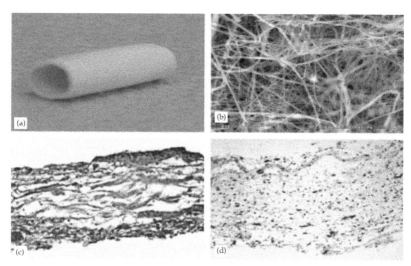

FIGURE 2.3 Characterization of the electrospun scaffold. (a) Electrospun nanofiber scaffold before crosslinking: collagen, elastin, and PLGA composite. (b) SEM image of electrospun nanofiber scaffold before crosslinking at 1800X magnification. (c) Immunohistochemical analyzes using antibodies specific to collagen type I in the scaffold. (d) The electrospun scaffold with 15% elastin demonstrated a uniform elastin matrix throughout the vascular scaffold wall. Reprinted from Biomaterials, Volume 27, Issue 7, Joel Stitzel, Jie Liu, Sang Jin Lee, Makoto Komura, Joel Berry, Shay Soker, Grace Lim, Mark Van Dyke, Richard Czerw, Controlled fabrication of a biological vascular substitute, 1088–1094 2006, with permission from Elsevier.

2.5   BIO-HYBRID VASCULAR GRAFTS BASED ON GELATIN/ SYNTHETIC POLYMERS

Gelatin is the denatured form of collagen with many of the same attractive qualities such as biological origin, biocompatibility, biodegradability and commercially available at the low cost, all owing to its popularity as a bio-polymer [17]. Since the very similar behavior as the collagen, gelatin has been used as a biopolymer for the vascular tissue engineering application. Electrospinning of gelatin into nano-micro scale fibers (50–500 nm) and a comparison of its mechanical characteristics with collagen and elastin were reported elsewhere [16]. Results have shown that the gelatin fibers have higher tensile modulus than collagen fibers, but have similar tensile strength and ultimate elongation [16]. This data ensures that gelatin can be electrospun in nano-micro scale fibers when proper solvent and concentration were used. As in the case of other biopolymers, gelatin fibers once implanted in vivo, are prone to enzymatic degradation. Consequently, in vascular graft applications, a scaffold made of only gelatin may not be sufficient enough to provide adequate mechanical strength; therefore it is more commonly incorporated with other synthetic biodegradable polymers.

In order to create a hybrid scaffold, gelatin was grafted onto electrospun PCL nanofibers, then ECs were seeded on the scaffold for cell proliferation studies [58]. First, PCL nanofibers were treated with the air plasma to introduce–COOH groups on the surface of the fibers, then gelatin molecule grafting was done. One of the reasons to apply gelatin grafting is to make the PCL fiber surface more hydrophilic, so the cell can attached easily [58]. The grafting of gelatin significantly improved the ECs cell attachment and the growth compared to only PCL fibers [58]. The same group had also tried to electrospun the gelatin fibers and also coelectrospun with 50:50 w/v solution of PCL [17]. The mechanical data (Table 2.2) evaluation showed that gelatin/PCL fibers have less tensile strength compared to PCL fibers, however, the elongation and deformation properties were drastically improved [17].

Synthetic polymers such as PCL have a disadvantage because they lack cell-binding sites unlike natural polymers. Therefore, combining the synthetic polymer with natural polymers (ECM proteins) will significantly improve the cell adhesion characteristic of the scaffold. One of the ways to incorporate natural polymer with the synthetic polymer is grafting. Previously, researchers have used grafting technique to attach the gelatin to the electrospun fibers. However, there are limitations associated with this technique, for instance, glutaraldehyde, a cross-linking agent, can greatly reduce the pore size of the scaffold and can be cytotoxic [59]. Therefore, scientists preferred to coelectrospun the gelatin and other synthetic biodegradable polymer. A comparative study using cospun collagen/gelatin/ PCL scaffold and gelatin/PCL scaffold [60] showed that the average fiber diameter was increased with the increasing amount of PCL in the PCL/ gelatin composition. In contrast to the fiber size, the pore size decreased with the increasing synthetic polymer concentration. The combination of 10% PCL with 10% gelatin has significantly improved the mechanical tensile strength compare to PCL/collagen or PCL/elastin.

PLCL electrospun fibers are usually very elastic and soft. On the other hand, pure gelatin electrospun fibers exhibited hard and brittle behavior. Therefore, a combination of the PLCL and gelatin solution can provide appropriate tensile mechanical strength and elongation properties. Lee et al. have blended poly (L-lactide-co ε-caprolactone) (PLCL) (50:50) co-polymer with gelatin to coelectrospun fibers [18]. Figure 2.4 illustrates clearly the proliferation of NIH-3T3 cells on the electrospun scaffolds made from PLCL [18]. The porosity of the scaffold increased as the gelatin concentration increased. PGA is a biodegradable polymer and currently been used in various medical applications. Scaffolds comprised of only PGA have been found to be far too brittle to withstand mechanical testing. The combination of PGA and 30% gelatin scaffold has shown very soft and flexible characteristics [20]. The area between the fibers also increased allowing for greater porosity as the concentration of the gelatin increased. The Young's modulus and elongation were found to be 32 MPa and 32% [20], respectively (Table 2.3). The HUVECs growth and proliferation was significantly higher on PGA and 30% gelatin scaffolds [20]. These results suggest that the gelatin had improved the cell attachment and the mechanical properties of the scaffold when compare with only PGA scaffold.

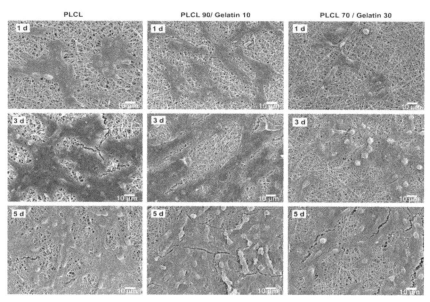

FIGURE 2.4 SEM images of NIH-3T3 cells proliferated on various PLCL/gelatin blended fiber sheets for 5 days. Reprinted from Biomaterials, Volume 29, Issue 12, Jongman Lee, Giyoong Tae, Young Ha Kim, In Su Park, Sang-Heon Kim, Soo Hyun Kim, The effect of gelatin incorporation into electrospun poly (L-Lactide-co-caprolactone) fibers on mechanical properties and cytocompatibility, 1872–1879, 2008 with permission from Elsevier.

TABLE 2.3   Mechanical Properties of Different Bio-Hybrid Compositions Used In Vascular Grafts

Scaffold Composition	Diameter of the fibers	Tensile Strength (MPa)	Tensile Modulus (MPa)	Strain at break (%)	Reference
Collagen type I	100–730 nm	1.5 ± 0.2	52.3 ± 5.2	–	[11]
PCL/collagen (wet)	520 nm	4.0±0.4	2.7 ± 1.2	140 ± 13	[12]
PCL/collagen I/collagen III	–	18	7.79	–	[13]
P(LLA- CL)/collagen I	470 ± 130 nm	6.27 ± 1.38	43.99 ± 4.04	176 ± 49	[14]
PDO/collagen I		4.6–6.7	7.6–18.0	56.5–186.4	[15]
Gelatin	460 ± 148–517 ± 241 nm	8–12	426 ± 39	80–100	[16]
Gelatin/PCL	–	1.29	30.8	1.38	[17]
PLCL/gelatin (50:50)	200 nm–1.3 μm	5.1 ± 0.8	340 ±8.5	86 ± 19	[18]
PLGA/gelatin/elastin	317 ± 46 nm	130 ± 7	770 ± 131	–	[19]
PGA/gelatin (30%)	863.96 ± 265.09 nm	–	32	32	[20]
Polyglyconate/gelatin/elastin	200–400 nm	2.71 ± 0.2	20.4 ± 3	140 ± 3	[21]
Polydioxanone/gelatin/elastin	300–500 nm	1.77 ± 0.2	5.74 ± 3	75.08 ±10	[22]
Fibrinogen	80–700 nm	2	80	–	[23]
Fibrinogen/ PLLA-CL (20:80)	305 ± 78 nm	10.1 ± 0.4	–	118.8 ± 12.3	[24]
PLGA/fibrin	50 nm–4 μm	0.651 ± 0.24	–	105 ± 10	[25]
Bacterial Cellulose		311 ± 29	–	–	[26]
B. mori Silk	9.3 ± 0.3 μm	740	10000	20	[27]
B. mori silk/ PEO	–	2.42±0.48	2.45±0.47	–	[28]
Silk fibroin/gelatin	143 ± 36 nm	1.12 ± 0.11	–	30.55 ± 3.46	[29]
PLA/SF – gelatin	–	1.28 ± 0.21	–	41.11 ± 2.17	[29]
Collagen/chitosan/ PLLA-CL (25:5:75)	409 ± 120 nm	16.9 ± 2.9	10.3 ± 1.1	112 ± 11	[30]
PEUU/PMBU (15%)	525±162 nm	7 ± 1	7 ± 3	342 ± 43	[31]

In search of a novel hybrid scaffolds, the combination of gelatin blends with other biopolymers such as elastin and synthetic polymers have been considered. Lelkes' group coelectrospun blends of PLGA, gelatin, and elastin to create a scaffold [19]. They found that by applying appropriate ratio of PLGA and gelatin, they could be able to control a fiber diameter and mechanical properties of the scaffold (Table 2.3). The scaffold showed promising in vitro cell attachment and support for ECs and SMCs as evident from the morphology and the cytoskeletal spreading. The ECs layer formed was also found to be nonthrombogenic and functional, which is an important and critical quality needed to be a successful vascular graft. Another interesting study has been done by using hydrogels for the vascular tissue engineering application. Hydrogels are appealing biomaterials for scaffold because they mimic the ECM of many tissues [61]. However, the hydrogel lacks the mechanical properties required. Therefore, Liu et al. studied hydrogels based on the interpenetrating polymer network (IPN) of gelatin and dextran bi-functionalized with methacrylate (MA) and aldehyde (AD) (Dex-MA-AD) [62] to improve mechanical properties and cell-encapsulation. As a result, the compressive modulus of elasticity was improved and the hydrogel also supported ECs dispersion throughout the hydrogel.

2.6   BIO-HYBRID VASCULAR GRAFTS BASED ON FIBRIN/ FIBRINOGEN

Fibrin and fibrinogen play an important role in blood coagulation. In addition, they also participate in fibrinolysis, cellular and matrix interaction, inflammation, wound healing, and neoplasia [63]. Fibrinogen is made of set of three polypeptide chains that are joined through disulfide bonds between N and E terminals. These domains play a vital role for the activation of fibrinogen into fibrin in the blood-clotting scenario. Cutaneous wound healing is the process of forming new ECM, cellular infiltration, and tissue formation, since fibrin plays a crucial role in the process, it may contribute similarly in the formation of the new blood vessels. Conversion of soluble fibrinogen to insoluble fibrin polymeric structure will present totally different mechanical characteristics. Fibrin provides a crucial role of protecting the underlying tissue structure while simultaneously promote migration and proliferation of the endothelial and stromal cells [64]. This characteristic of fibrin attracted many scientists to incorporate or use the fibrin in the vascular tissue regeneration. In addition, fibrin has showed higher affinity for vascular endothelial growth factor (VEGF) and fibrinogen growth factor (FGFs) [65]. Figure 2.5 illustrates the scaffold made of fibrin and poly lactide for vascular tissue engineering application implanted as an end-to-end anastomosis (arrowheads) to the common carotid artery [66]. PCL scaffold (film) fabricated by solvent casting method with a pore size of 5–200 μm was coated with a fibrin composite to develop a biomimetic scaffold for vascular tissue engineering applications [67]. Human umbilical vein endothelial cells (HUVECS) cultured on the hybrid scaffold formed a continuous EC layer within 15 days. Further, no adverse effect cell growth was noticed.

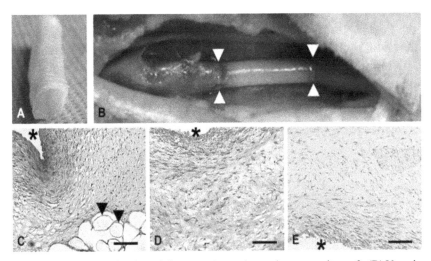

FIGURE 2.5 (A) Pre-implanted tissue-engineered vascular composite graft; (B) Vascular composite grafts were implanted as an end-to-end anastomosis (arrowheads) to the common carotid artery; (C) H&E staining of preimplanted grafts revealed a homogenous cell distribution within the fibrin-based cell carrier, which was supported by P(L/D)LA 96/4 fibers (arrowheads); (D) Gomori's trichrome staining illustrated abundant collagen (blue) within the preimplanted grafts; (E) absence of calcific deposits prior to implantation was evidenced by van Kossa stain (black). Images CeE (graft mid-section). Graft lumen is indicated by *. Scale bars: 200 μm. Reprinted from Biomaterials, Volume 31, Issue 17, Sabine Koch, Thomas C. Flanagan, Joerg S. Sachweh, Fadwa Tanios, Heike Schnoering, Thorsten Deichmann, Ville Ella, Fibrin-polylactide-based tissue-engineered vascular graft in the arterial circulation, 4731–4739, 2010, with permission from Elsevier.

Electrospinning of fibrinogen has also been reported. Bowlin's group has attempted to electrospin fibrinogen and create nanofibers [23] from lyophilized human and bovine fibrinogen fraction I from plasma. The fiber diameter was directly related to the concentration of the fibrinogen content; the higher concentration resulted in increased diameter. When the concentration of the fibrinogen gradually increased from 0.083 g/mL to 0.167 g/mL in HFP/MEM solution, the fiber diameter was increased from 80 ± 10, to 700 ± 110 nm [23]. Since, the fiber diameter plays a critical role in the mechanical properties and porosity of the scaffold, ability to produce a broad range of fiber diameter may provide high flexibility in the application. The orientation of the fibers will also affect the tensile properties of the scaffold; therefore having a control on the fibers orientation during electrospinning can be beneficial. The choice of proper synthetic polymer plays a vital role in the tensile properties, degradation rate, and elasticity of the final scaffold. The studies on mixed fibrinogen/PLLA-CL [24] and culturing of mouse fibroblast onto the scaffold suggests that by combining the fibrinogen to the PLLA-CL, a hybrid scaffold resulted in has more hydrophilic and bioactive nature in comparison to 100% PLLA-CL scaffold.

2.7 BIO-HYBRID VASCULAR GRAFTS BASED ON BACTERIAL CELLULOSE

The use of bacterial cellulose (BC) as a biopolymer in the fabrication of small diameter vascular grafts has also been explored. Previous applications for this material have included dermal substitutes and micro vessel endoprosthesis. Preliminary studies have also suggested BC may be a suitable scaffold material for cartilaginous applications [68]. Cellulose has been championed as versatile biopolymer exhibiting not only the biocompatibility, hydrophilicity, biodegradability, moldability, and mechanical tunability necessary for tissue regeneration, but also, following a general trend, the ability to be a sustainable "green" material. Through a culture medium, *Acetobacter xylinum* bacteria emits long nonaggregated polysaccharide chain [69]. Figure 2.6 illustrates the bacterial cellulose network [69] Vascular grafts derived from this material has been fabricated using a patented matrix reservoir technology [70]. This method allows hollow tubes of various length, inner diameter, and wall thickness to be molded from a culture medium [71].

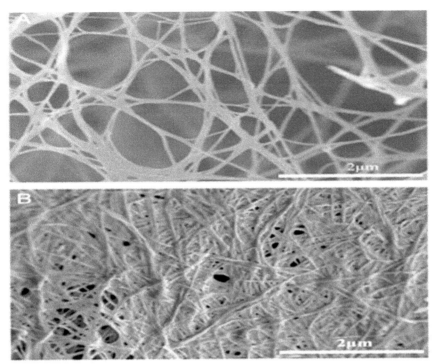

FIGURE 2.6 High resolution SEM images of the BC network. The early network (a) will eventually become a thin dense layer of the inner BC tube region (b). Reprinted from Materials Science and Engineering: C, Volume 31, Issue 1, Henrik Bäckdahl, Bo Risberg, Paul Gatenholm, Observations on bacterial cellulose tube formation for application as vascular graft, 14–21, 2011, with permission from Elsevier.

Schumann et al. used small diameter vascular grafts (diameter 1.0–3.7, length 5.0–10.0, and wall thickness 0.7 mm) comprised of bacterial cellulose to study long-term in vivo replacement of the carotid artery in rats and in pigs [72]. Excision of the BC graft after one year in the low-pressure rat test subject showed BC integration within the formation of neointima and active fibroblasts [72]. Examination of the lumen via immune histo chemical labeling (CD-31) and SEM imaging showed a comprehensive layer of endothelial cells without any signs of thrombosis [72]. Further in vivo testing in high-pressure, pig models showed only 1 of 8 grafts occluded after 90 days. There was no observable change to the graft's inner diameter and overall formation after excision.

Zahedmanesh et al. studied the mechanical properties and vascular cell adhesion and proliferation of bacterial cellulose vascular grafts of 4 mm diameter [26]. Ultimate tensile strength determined from initial failure of sample was 311 ± 29 kPa. No variation in stress-strain response was observed in axial or circumferential testing with a stress range between 0.8–1.3 kPa. Compliance tests resulted in a mean measurement of $4.27 \pm 1.51 \ 10^{-2}$% per millimeter of mercury with a mean luminal pressure range of 30–120 mmHg [26]. Bovine SMCs were cultured to show penetrated growth (more than 100 μm) into the outer layer. A subsequent cell adhesion using endothelial cells showed adhesion and proliferation of cells on the luminal side, which was precoated with fibronectin.

Blending of heparin and bacterial cellulose (Hep-BC) has been carried out to improve blood compatibility. Heparin is known for processing anticoagulant sulfate groups [73]. Thrombosis is a reoccurring issue in synthetic vascular engineering. Heparin has already been shown to aid in vascular constructs as a viable surface component [74] with anticoagulant properties. 0.1% w/v Heparin was combined with BC medium and spun into a blended BC-Heparin nanofibers. Scanning electron microscopy (SEM) images revealed no change in morphological structure of the BC fibers when blended with Heparin. Chemical analyzes indicated the presence of sulfate groups necessary for anticoagulant properties. A decrease in crystallinity in Hep-BC hybrid samples suggests a difference in the atomic structure due to the cosynthesis technique applied [73].

2.8   BIO-HYBRID VASCULAR GRAFT BASED ON SILK FIBROIN/ SYNTHETIC POLYMERS

Silk is a natural protein produced by different species of spiders and also by several worms of the order Lepidoptera, which include mites, butterflies, and moths [75]. It is generally composed of the β sheets structures. The β sheets are made of hydrophobic domains consist of small amino acid sequences to provide structural support. In tissue engineering field, silk from silkworms and orb waving spider been frequently explored because of their biocompatibility and tunable structural properties [76]. The small amino acid sequences are the most attractive quality of the silk fiber as it provides the mechanical integrity, ability for cells to migrate and proliferate, tunable degradation rate, and potential for surface modification.

Since the silk can be produce by different species, the structural properties can also be widely different due to various amino acid sequences and environmental effects.

Non-woven silk fibrin mats became an attractive biomaterial for tissue engineering because of the higher surface area and rougher topography for cell attachment [75]. These mats are formed by dissolving the natural silk fibers into aqueous solution and electrospun into fibers in nano to micro range. It is easy to perform surface modifications for attracting a specific cell lines or induce a particular signaling pathway. For example, the silk fibers surface can be modified by incorporating the RGDS sequence and has been shown to help enhance the cellular affinity towards integrin who recognize the RGDS sequences. Usually, the silk degrades in vivo by the action of proteases. Further, the rate of degradation depends on cross-linking of the biopolymers. The biopolymer with the higher cross-linking may reduce the degradation rate.

Kaplan's group has fabricated a small diameter tubular scaffold based on silk protein by electrospinning and compared the mechanical properties and cell viability with PTFE scaffold, a highly used material to make a large diameter scaffold [77]. In vitro characterization of cell attachment, proliferation and platelet attachment after seeding the scaffolds with HU-VECs, and human umbilical vein smooth muscle cells (HUVSMCs). Further, the silk scaffold was implanted into the rat aorta via end to end anastomosis model. Silk fibroin scaffold showed higher anti thrombogenicity compare to PTFE scaffold. The silk scaffolds showed better patency after 4 weeks of period in comparison with PTFE scaffold (few days of implantation) [77]. In another study, Marelli et al. had electrospun vascular graft with 6-mm diameter from silk fibrin [78]. They observed the cell attachment and proliferation seeded NIH 3Y3 fibroblast cells for 24, 48, 72 h and 5, 7 days period. Kaplan's group has also studied the use of silk-based electrospun scaffold for the vascular graft application [28]. The scaffold was prepare for the small diameter (<6-mm) vascular graft. In order to electrospin silk, the silk solution was mixed with the PEO to make the solution more viscous and spinnable. The spun scaffold exhibited promising tensile properties (Table 2.2) with a burst pressure ranging from 704 to 919 mmHg). Human aortic endothelial cells (HAEC) seeded on the scaffold were found to be attached, spread and grew on the electrospun mats with the production of cell-ECM [28]. Figure 2.7 shows the electrospun silk fibers with PEO nonextracted and extracted to visualize the silk scaffold morphology [79].

Liu et al. have used the biocompatible properties of silk to improve hemocompatibility and endothelialization of vascular graft made of PLGA [80]. One of the biggest difficulties is to achieve a consistent endothelial cell layer in the lumen. Loosely attached ECs will easily separate from the lumen wall and washed away in vivo condition. Therefore, the sulfated silk fibroins were prepared to improve antithrombogenicity and enhance cytocompatibility of the scaffold made from silk fibroin. Liu et al. have also electrospun sulfated silk nanofibers for vascular tissue engineering application [81]. Sulfated silk fibers were created by standard electrospinning method using chlorosulfonic acid and pyridine. ECs and SMCs seeded onto the scaffold in vitro condition indicated that the sulfated silk fibroin has a better affinity for cell than pure silk fibroin [81].

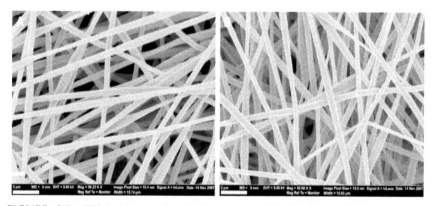

FIGURE 2.7 SEM micrographs of electrospun silk fibroin scaffolds: (A) PEO nonextracted scaffold; (B) PEO extracted scaffold. Scale bars are 2 μm. Reprinted from Biomaterials Volume 29, Issue, Xiaohui Zhang, Cassandra B. Baughman, David L. Kaplan, In vitro evaluation of electrospun silk fibroin scaffolds for vascular cell growth, 2217–2227, 2008, with permission from Elsevier.

In the human body ECM has a general defined arrangement of the protein fibers depending on the tissue type. Bowlin's group has fabricated highly aligned and random electrospun PDO/silk and PCL/Silk fibroin scaffold for vascular regeneration [82]. In order to achieve aligned and random fibers of PDO/Silk and PCL/silk, the RPM of the mandrel were ranged from 500 to 8000. Wang et al. have fabricated another electrospun vascular graft made from PLA/Silk/gelatin fibers for small diameter graft applications [29]. The composite graft was constructed in such a way that the outer layer comprised of PLA and inner layer comprised of silk-gelatin. The biomechanical properties (Table 2.2), the effect of wall thickness of the scaffold on the suture retention, and burst pressure properties were assessed. As the wall thickness of the scaffold increased the suture retention and burst pressure properties were improved. For instance, a scaffold with the 0.3 mm wall thickness has approximate 61 kPa burst pressure in comparison a scaffold with 0.7 mm thickness has approximate 172 kPa burst pressure [29]. They also noticed that cell growth and proliferation was higher on the PLA/silk-gelatin scaffold than PLA/silk scaffold.

2.9 MULTI-LAYERED BIO-HYBRID VASCULAR GRAFTS

Generally, sequential spinning technique been used to generate a multiple layers in the electrospun tubular graft fabrication. Multilayering by using electrospinning was first developed by the Kidoaki et al. using type I collagen, polyurethane, PEO, and gelatin [83]. Since then, many researchers have made bi and tri layer scaffold to mimic the native layered structure of blood vessels. A bi-layered structure was construct by first electrospinning wet PCL fibers to mimic the lumen layer, and then another electrospun polyurethane (PU) fibers to generate outer layer [84]. This will provide two different mechanical characteristics and degradation rate to each layer of the bi-layered vascular graft. Electrospinning technique also combined with a freeze-drying technique to create bi-layered scaffolds for the vascular tissue engineering applications. PLGA was electrospun to create an out layer and then freeze dried collagen type I was incorporated as inner layer to improve both biocompatibility and mechanical properties of constructed bi-layered conduit [85].

In order to mimic the native layer structure of blood vessels, Thomas' group has designed functionally graded tri-layered electrospun scaffolds with the blend of polymers and vascular proteins (collagen, gelatin and elastin) [22, 86]. The inner layer of the vascular conduits was mainly composed of the natural polymer to achieve better biocompatibility and non-thrombogenic behavior. The outer layer (media like) was composed of blend of natural and synthetic polymer. Finally, the tubular scaffold had a spatially designed tri layered structure consisting of higher nature protein content inner side and gradually decreases on outer side and simultaneously synthetic polymer content increase from inner side to outer side. Electrospinning technique was used to create the multilayered scaffold made of polyglyconate, gelatin and elastin with tunable mechanical properties [21]. The different amount of polymer, gelatin and elastin were used in each layer to create a graded matrix composition. The polyglyconate or Maxon has been used for surgical suture materials; therefore it is flexible, biocompatible with favorable degradation rate. The ratio of the three different components was optimized to obtain optimum mechanical properties. The porosity was also increased when gelatin and elastin was incorporated compared with only polyglyconate scaffolds. Even in the hydrated condition, the scaffold made from the combination of polygyconate, gelatin, and elastin maintained the mechanical properties comparable to native femoral artery (Table 2.2). Another tri layered electrospun tubular conduit was also made from using polydioxanone, elastin and gelatin with different ratio (Table 2.2) [22]. The group has further tried to incorporated PCL, polyglecaprone (PGC), gelatin, and elastin to create a nano fibrous electrospun scaffold and incorporated human aortic endothelial cells (HAECs) to promote the endothelialization on the scaffold [86b]. A new tri layered vascular graft structure was obtained by using silk fibroin, collagen, elastin and PCL by Bowlin's group [87]. Dynamic compliance and the burst pressure data, 2.5%/100 mmHg and 1614–3500 mmHg, respectively, were comparable to native saphenous vein. Furthermore, there was no significant difference noticed in these mechanical properties after 4 weeks of degradation study.

2.10 SUMMARY

Patients exhibiting altered anatomy or disease may need to use a vascular graft made of biocompatible materials for the repair of the damaged tissue during cardiovascular surgery [88]. To overcome the limitations posed by the autograft, allograft, and synthetic materials; there is a need to create a new generation of hybrid materials which will have proper mechanical and structure properties, as well as appropriate surface characterization for cell attachment, proliferation, and regeneration [41]. Mimicking the ECM structural properties remains the ideal goal; hence scientists are working on multiple components of ECM especially ECM proteins to create an ideal scaffold. The most common proteins found in ECM such as collagen (gelatin), elastin, fibrin, and other biopolymers such as silk, bacterial cellulose have been used in vascular tissue engineering. Even though these proteins together provide a structural stability for vascular ECM, they are not quite capable of making an ideal vascular graft alone. While they have promising biocompatibility and cell-growth characteristics they are also prone to early enzymatic degradation and lacking sufficient bio-mechanical properties. On the other hand, the vascular scaffolds made of synthetic polymer such as PLA, PCL, PLGA or PDO exhibit greater mechanical properties and able to provide structural and dimensional stability. Due to the decreased bioactivity of these synthetic polymers, a combination of biopolymers and synthetic polymers to create vascular grafts is essential. The ideal tissue engineered vascular graft is the one which promotes neo-tissue formation without promoting inflammation, and which degrades as the neo-tissue matures.

While there have been many techniques used to create a small diameter vascular graft, electrospinning remains the preferred method to produce native ECM mimicking scaffolds. The electrospun scaffolds have a larger surface area, so the cell can easily adhere, proliferate and eventually secrete their own ECM. The porosity of the electrospun scaffolds is much higher than those made from other methods. This quality is critical for cellular penetration and ECM spreading into three-dimension space. However, generally observed small pore size of electrospun scaffold is a concern for cellular penetration. Continuous research is still underway to obtain the ideal blends of synthetic and natural polymer required for proper biomechanical properties such as tensile strength, elastic of modulus, cyclic fatigue, burst pressure, suture retention, and degradation and growth enhancing characteristics.

Despite over 50 years of research in the area of vascular tissue engineering, there is no suitable tissue engineered vascular graft for small diameter blood vessel (<6 mm) substitution available for patient care. Still, native vein substitution remains the gold standard method for the revascularization procedures [89]. Current, synthetic or hybrid vascular grafts has been associated with four major issues: (i) thrombosis due to lack of endothelium cell layer; (ii) restenosis caused by chronic inflammatory response; (iii) susceptibility to infection 4) lack of mechanical properties to sustain in vivo environment [44, 90]. Various approaches to incorporate collagen and fibrin and seeding cells to create autologous arterial grafts have been developed, but they failed to have clinically useful burst pressure characteristics [91]. Recent reports on the use of bio-hybrid scaffolds systems for vascular tissue regeneration are encouraging. Electrospun recombinant human tropoelastin (rTE) [92], coelectrospun (PLLA–CL), collagen and chitosan [30], heparin conjugated with PCL/gelatin electrospun scaffold to provide control release of platelet derived growth factor BB (PDGF-BB) [93], cospun biodegradable poly (ester urethane) urea (PEUU) and poly(2-methacryloyloxyethyl phosphorylcholine-comethacryloyloxyethyl butylurethane) (PMBU) a phospholipid with a non thrombogenic behavior [31] were fabricated as bio-hybrid scaffolds for vascular regeneration. The authors have focused on the development of functionally graded bio-hybrid scaffolds of polymers and collagenous biomatrix for tuning the biomechanical properties and cellular growth combined with controlled release of growth factors or antithrombotic agents for small- diameter vascular graft regeneration. The polymer-rich outer layer is expected to prevent overstretch or rupture and the softer biomatrix-rich is expected to be conductive enough for attracting ECs and progenitor cells and subsequent endothelialization. In conclusion, biopolymers continue to contribute significantly to create an ideal bio-hybrid vascular graft. The source and the concentration of biopolymers will definitely play a vital role in the vascular grafts characterizations. Proper selection of materials, chemical composition (ratio of biopolymers to synthetic bioresorbable polymers) and design of tubular 3D conduits with biomimetic nano-features together with cells and growth factors have a massive impact on the success of vascular tissue engineering.

2.11 ACKNOWLEDGMENTS

We acknowledge the support by Prof. Yogesh Vohra and UAB Center for Nanoscale Materials and Biointegration (CNMB). The authors apologize to those authors whose published works may have been relevant to this review but was not cited due to perceived lack of fit or due to space limitations.

KEYWORDS

- **Bio-hybrid material**
- **Biopolymers**
- **Electrospinning**
- **Mechanical properties**
- **Vascular tissue graft**

REFERENCES

1. Rosamond, W., Flegal, K., Friday, G., Furie, K., Go, A., Greenlund, K., Haase, N., Ho, M., Howard, V., Kissela, B., Kittner, S., Lloyd-Jones, D., McDermott, M., Meigs, J., Moy, C., Nichol, G., O'Donnell, C. J., Roger, V., Rumsfeld, J., Sorlie, P., Steinberger, J., Thom, T., Wasserthiel-Smoller, S., & Hong, Y. (2007). Heart disease and stroke statistics 2007 update: a report from the American Heart Association Statistics Committee and Stroke Statistics Subcommittee. Circulation, 115(5), 69–171.
2. Stegemann, J. P., Kaszuba, S. N., & Rowe, S. L. (2007). Review: advances in vascular tissue engineering using protein-based biomaterials. Tissue engineering, 13(11), 2601–13.
3. Soletti, L., Hong, Y., Guan, J., Stankus, J. J., El-Kurdi, M. S., Wagner, W. R., & Vorp, D. A. (2010). A bilayered elastomeric scaffold for tissue engineering of small diameter vascular grafts. Acta biomaterialia, 6(1), 110–122.
4. (a) Lloyd-Jones, D., Adams, R. J., Brown, T. M., Carnethon, M., Dai, S., De Simone, G., Ferguson, T. B., Ford, E., Furie, K., & Gillespie, C. (2010). Heart disease and stroke statistics 2010 update. Circulation, 121(7), e46–e215.
(b) Thomas, V., Zhang, X., Catledge, S. A., & Vohra, Y. K. (2007). Functionally graded electrospun scaffolds with tunable mechanical properties for vascular tissue regeneration. Biomedical Materials, 2(4), 224.
5. Dahl, S. L., Kypson, A. P., Lawson, J. H., Blum, J. L., Strader, J. T., Li, Y., Manson, R. J., Tente, W. E., DiBernardo, L., & Hensley, M. T. (2011). Readily available tissue-engineered vascular grafts. Sci Transl Med, 3(68), 68ra69.
6. (a) Collaboration, A. T. (1994). Collaborative overview of randomized trials of anti-platelet therapy Prevention of death, myocardial infarction, and stroke by prolonged antiplatelet therapy in various categories of patients. BMJ, 308(6921), 81–106.
(b) Deterling, R. A., & Bhonslay, S. B. (1955). An evaluation of synthetic materials and fabrics suitable for blood vessel replacement. Surgery, 38(1), 71–91.
(c) Zdrahala, R. J. (1996). Small Caliber Vascular Grafts. Part I: State of the Art. Journal of Biomaterials Applications, 10(4), 309–329.
7. (a) Bergan, J. J., Veith, F. J., Bernhard, V. M., Yao, J. S., Flinn, W. R., Gupta, S. K., Scher, L. A., Samson, R. H., & Towne, J. B. (1982). Randomization of autogenous

vein and polytetrafluorethylene grafts in femoral-distal reconstruction. Surgery 92(6), 921–930.

(b) Hess, F. (1985). History of (MICRO) vascular surgery and the development of small-caliber blood vessel prostheses (with some notes on patency rates and reendothelialization). Microsurgery, 6(2), 59–69.

(c) Pevec, W. C., Darling, R. C., L'Italien, G. J., & Abbott, W. M. (1992). Femoropopliteal reconstruction with knitted, nonvelour dacron versus expanded polytetrafluoroethylene. Journal of Vascular Surgery, 16(1), 60–65.

8. Yamada, H., & Evans, F. G. (1970). Strength of biological materials. Williams & Wilkins Baltimore Vol. 1000.

9. Kurane, A., Simionescu, D. T., & Vyavahare, N. R. (2007). In vivo cellular repopulation of tubular elastin scaffolds mediated by basic fibroblast growth factor. Biomaterials, 28(18), 2830–2838.

10. Valenta, J., & Jaroslav, V. (1993). Clinical aspects of biomedicine. Biomechanics, 2, 142–179.

11. Matthews, J. A., Wnek, G. E., Simpson, D. G., & Bowlin, G. L. (2002). Electrospinning of collagen nanofibers. Biomacromolecules, 3(2), 232–238.

12. Lee, S. J., Liu, J., Oh, S. H., Soker, S., Atala, A., & Yoo, J. J. (2008). Development of a composite vascular scaffolding system that withstands physiological vascular conditions. Biomaterials, 29(19), 2891–2898.

13. Venugopal, J., Zhang, Y., & Ramakrishna, S. (2005). Fabrication of modified and functionalized polycaprolactone nanofiber scaffolds for vascular tissue engineering. Nanotechnology, 16(10), 2138.

14. He, W., Yong, T., Teo, W. E., Ma, Z., & Ramakrishna, S. (2005). Fabrication and endothelialization of collagen-blended biodegradable polymer nanofibers: potential vascular graft for blood vessel tissue engineering. Tissue engineering, 11(9–10), 1574–1588.

15. Barnes, C., Sell, S., Knapp, D., Walpoth, B., Brand, D., & Bowlin, G. (2007). Preliminary investigation of electrospun collagen and polydioxanone for vascular tissue engineering applications. International Journal of Electrospun Nanofibers and Applications, 1(1), 73–87.

16. Li, M., Mondrinos, M. J., Gandhi, M. R., Ko, F. K., Weiss, A. S., & Lelkes, P. I. (2005). Electrospun protein fibers as matrices for tissue engineering. Biomaterials, 26(30), 5999–6008.

17. Zhang, Y., Ouyang, H., Lim, C. T., Ramakrishna, S., & Huang, Z. M. (2005). Electrospinning of gelatin fibers and gelatin/PCL composite fibrous scaffolds. Journal of Biomedical Materials Research Part B: Applied Biomaterials, 72B (1), 156–165.

18. Lee, J., Tae, G., Kim, Y. H., Park, I. S., Kim, S. H., & Kim, S. H. (2008). The effect of gelatin incorporation into electrospun poly(l-lactide-co□-caprolactone) fibers on mechanical properties and cytocompatibility. Biomaterials, 29(12), 1872–1879.

19. Han, J., Lazarovici, P., Pomerantz, C., Chen, X., Wei, Y., & Lelkes, P. I. (2010). Co-electrospun blends of PLGA, gelatin, and elastin as potential nonthrombogenic scaffolds for vascular tissue engineering. Biomacromolecules, 12(2), 399–408.

20. Hajiali, H., Shahgasempour, S., Naimi-Jamal, M. R., & Peirovi, H. (2011). Electrospun PGA/gelatin nanofibrous scaffolds and their potential application in vascular tissue engineering. International Journal of Nanomedicine, 6, 2133.

21. Thomas, V., Zhang, X., Catledge, S. A., & Vohra, Y. K. (2007). Functionally graded electrospun scaffolds with tunable mechanical properties for vascular tissue regeneration. Biomed Mater, 2(4), 224–32.

22. Thomas, V., Zhang, X., & Vohra, Y. K. (2009). A biomimetic tubular scaffold with spatially designed nanofibers of protein/PDS bio-blends. Biotechnology and bioengineering, 104(5), 1025–33.

23. Wnek, G. E., Carr, M. E., Simpson, D. G., & Bowlin, G. L. (2003). Electrospinning of nanofiber fibrinogen structures. Nano Letters, 3(2), 213–216.

24. He, C., Xu, X., Zhang, F., Cao, L., Feng, W., Wang, H., & Mo, X. (2011). Fabrication of fibrinogen/P (LLA-CL) hybrid nanofibrous scaffold for potential soft tissue engineering applications. Journal of Biomedical Materials Research Part A, 97(3), 339–347.

25. Chennazhi, K. P., Perumcherry, S., Menon, D., & Nair, S. (2012). Fabrication of electrospun PLGA-Fibrin multiscale scaffold for myocardial regeneration. Tissue engineering, (ja).

26. Zahedmanesh, H., Mackle, J. N., Sellborn, A., Drotz, K., Bodin, A., Gatenholm, P., & Lally, C. (2011). Bacterial cellulose as a potential vascular graft: Mechanical characterization and constitutive model development. Journal of Biomedical Materials Research Part B: Applied Biomaterials, 97B (1), 105–113.

27. Pérez-Rigueiro, J., Viney, C., Llorca, J., & Elices, M. (2000). Mechanical properties of single-brin silkworm silk. Journal of Applied Polymer Science, 75(10), 1270–1277.

28. Soffer, L., Wang, X., Zhang, X., Kluge, J., Dorfmann, L., Kaplan, D. L., & Leisk, G. (2008). Silk-based electrospun tubular scaffolds for tissue-engineered vascular grafts. Journal of Biomaterials Science, Polymer Edition, 19(5), 653–664.

29. Wang, S., Zhang, Y., Yin, G., Wang, H., & Dong, Z. (2009). Electrospun polylactide/ silk fibroin–gelatin composite tubular scaffolds for small-diameter tissue engineering blood vessels. Journal of Applied Polymer Science, 113(4), 2675–2682.

30. Yin, A., Zhang, K., McClure, M. J., Huang, C., Wu, J., Fang, J., Mo, X., Bowlin, G. L., Al-Deyab, S. S., & El-Newehy, M. (2012) Electrospinning collagen/chitosan/ poly(L-lactic acid-co-caprolactone) to form a vascular graft: Mechanical and biological characterization. Journal of Biomedical Materials Research Part A.

31. Hong, Y., Ye, S. H., Nieponice, A., Soletti, L., Vorp, D. A., & Wagner, W. R. (2009). A small diameter, fibrous vascular conduit generated from a poly(ester urethane)urea and phospholipid polymer blend. Biomaterials, 30(13), 2457–2467.

32. Seliktar, D., Black, R. A., Vito, R. P., & Nerem, R. M. (2000). Dynamic mechanical conditioning of collagen-gel blood vessel constructs induces remodeling in vitro. Annals of biomedical engineering, 28(4), 351–362.

33. McPherson, T. B., & Badylak, S. F. (1998). Characterization of fibronectin derived from porcine small intestinal submucosa. Tissue engineering, 4(1), 75–83.

34. L'Heureux, N., Pâquet, S., Labbé, R., Germain, L., & Auger, F. A. (1998). A completely biological tissue-engineered human blood vessel. The FASEB Journal, 12(1), 47–56.

35. (a) McCullen, S. D., Ramaswamy, S., Clarke, L. I., & Gorga, R. E. (2009). Nanofibrous composites for tissue engineering applications. Wiley Interdisciplinary Reviews: Nanomedicine and Nanobiotechnology, 1(4), 369–390.

(b) Liu, W., Thomopoulos, S., & Xia, Y. (2011). Electrospun Nanofibers for Regenerative Medicine. Advanced Healthcare Materials.

(c) Mun, C. H., Jung, Y., Kim, S. H., Lee, S. H., Kim, H. C., Kwon, I. K., & Kim, S. H. (2012). Three-Dimensional Electrospun Poly (Lactide-Co-□-Caprolactone) for Small-Diameter Vascular Grafts. Tissue Engineering Part A.

36. Liao, S., Li, B., Ma, Z., Wei, H., Chan, C., & Ramakrishna, S. (2006). Biomimetic electrospun nanofibers for tissue regeneration. Biomed Mater, 1(3), R45–53.

37. (a) Thomas, V., Dean, D. R., & Vohra, Y. K. (2006). Nanostructured biomaterials for regenerative medicine. Current Nanoscience, 2(3), 155–177.

(b) Sell, S. A., McClure, M. J., Garg, K., Wolfe, P. S., & Bowlin, G. L. (2009). Electrospinning of collagen/biopolymers for regenerative medicine and cardiovascular tissue engineering. Advanced drug delivery reviews, 61(12), 1007–19.

(c) Liao, S., Ramakrishna, S., & Ramalingam, M. (2011). Development of Nanofiber Biomaterials and Stem Cells in Tissue Engineering. Journal of Biomaterials and Tissue Engineering, 1(2), 111–128.

(d) Collins, G., Federici, J., Imura, Y., & Catalani, L. H. (2012). Charge generation, charge transport, and residual charge in the electrospinning of polymers: A review of issues and complications. Journal of Applied Physics, 111(4), 044701–044701–18.

38. Veit, G., Kobbe, B., Keene, D. R., Paulsson, M., Koch, M., & Wagener, R. (2006). Collagen XXVIII, a novel von Willebrand factor A domain-containing protein with many imperfections in the collagenous domain. Journal of Biological Chemistry, 281(6), 3494–3504.

39. Pankajakshan, D., & Agrawal, D. K. (2010). Scaffolds in tissue engineering of blood vessels. Canadian Journal of Physiology & Pharmacology, 88(9), 855–873.

40. (a) Desai, T. A. (2000). Micro-and nanoscale structures for tissue engineering constructs. Medical engineering & physics, 22(9), 595.

(b) Wallace, D. G., & Rosenblatt, J. (2003). Collagen gel systems for sustained delivery and tissue engineering. Advanced drug delivery reviews, 55(12), 1631–1649.

41. Gomes, S., Leonor, I. B., Mano, J. F., Reis, R. L., & Kaplan, D. L. (2012). Natural and Genetically Engineered Proteins for Tissue Engineering. Progress in polymer science, 37(1), 1–17.

42. Weinberg, C., & Bell, E. (1986). A blood vessel model constructed from collagen and cultured vascular cells. Science, 231(4736), 397–400.

43. Stitzel, J. D., Pawlowski, K. J., Wnek, G. E., Simpson, D. G., & Bowlin, G. L. (2001). Arterial Smooth Muscle Cell Proliferation on a Novel Biomimicking, Biodegradable Vascular Graft Scaffold. Journal of Biomaterials Applications, 16(1), 22–33.

44. Veith, F. J., Gupta, S. K., Ascer, E., White-Flores, S., Samson, R. H., Scher, L. A., Towne, J. B., Bernhard, V. M., Bonier, P., & Flinn, W. R. (1986). Six-year prospective multicenter randomized comparison of autologous saphenous vein and expanded polytetrafluoroethylene grafts in infrainguinal arterial reconstructions. Journal of Vascular Surgery, 3(1), 104–114.

45. Buján, J., García-Honduvilla, N., & Bellón, J. M. (2004). Engineering conduits to resemble natural vascular tissue. Biotechnology and applied biochemistry, 39(1), 17–27.

46. He, W., Ma, Z., Yong, T., Teo, W. E., & Ramakrishna, S. (2005). Fabrication of collagen-coated biodegradable polymer nanofiber mesh and its potential for endothelial cells growth. Biomaterials 26(36), 7606–7615.

47. Wan, Y., Yang, J., Yang, J., Bei, J., & Wang, S. (2003). Cell adhesion on gaseous plasma modified poly (L-lactide) surface under shear stress field. Biomaterials, 24(21), 3757–3764.
48. Ju, Y. M., Choi, J. S., Atala, A., Yoo, J. J., & Lee, S. J. (2010). Bilayered scaffold for engineering cellularized blood vessels. Biomaterials, 31(15), 4313–4321.
49. Song, Y., Feijen, J., Grijpma, D., & Poot, A. (2011). Tissue engineering of small-diameter vascular grafts: A literature review. Clinical hemorheology and microcirculation, 49(1), 357–374.
50. Boland, E. D., Matthews, J. A., Pawlowski, K. J., Simpson, D. G., Wnek, G. E., & Bowlin, G. L. (2004). Electrospinning collagen and elastin: preliminary vascular tissue engineering. Front Biosci, 9(1422), C1432.
51. Leach, J. B., Wolinsky, J. B., Stone, P. J., & Wong, J. Y. (2005). Crosslinked α-elastin biomaterials: towards a processable elastin mimetic scaffold. Acta biomaterialia, 1(2), 155–164.
52. Ramalingam, M., Haidar, Z., Ramakrishna, S., Kobayashi, H., & Haikel, Y. (2012). Integrated Biomaterials in Tissue Engineering. Wiley-Scrivener Vol. 1.
53. Buttafoco, L., Kolkman, N. G., Engbers-Buijtenhuijs, P., Poot, A. A., Dijkstra, P. J., Vermes, I., & Feijen, J. (2006). Electrospinning of collagen and elastin for tissue engineering applications. Biomaterials, 27(5), 724–734.
54. McClure, M. J., Sell, S. A., Simpson, D. G., & Bowlin, G. L. (2009). Electrospun polydioxanone, elastin, and collagen vascular scaffolds uniaxial cyclic distension. Journal of Engineered Fibers and Fabrics, 4, 18–25.
55. Olde Damink, L., Dijkstra, P., Van Luyn, M., Van Wachem, P., Nieuwenhuis, P., & Feijen, J. (1996). Cross-linking of dermal sheep collagen using a water-soluble carbodiimide. Biomaterials, 17(8), 765–773.
56. Caves, J. M., Kumar, V. A., Martinez, A. W., Kim, J., Ripberger, C. M., Haller, C. A., & Chaikof, E. L. (2010). The use of microfiber composites of elastin-like protein matrix reinforced with synthetic collagen in the design of vascular grafts. Biomaterials, 31(27), 7175–7182.
57. Stitzel, J., Liu, J., Lee, S. J., Komura, M., Berry, J., Soker, S., Lim, G., Van Dyke, M., Czerw, R., Yoo, J. J., & Atala, A. (2006). Controlled fabrication of a biological vascular substitute. Biomaterials, 27(7), 1088–1094.
58. Ma, Z., He, W., Yong, T., & Ramakrishna, S. (2005). Grafting of gelatin on electrospun poly (caprolactone) nanofibers to improve endothelial cell spreading and proliferation and to control cell orientation. Tissue engineering 11(7–8), 1149–1158.
59. Van Wachem, P., Van Luyn, M., Damink, L., Dijkstra, P., Feijen, J., & Nieuwenhuis, P. (1994). Biocompatibility and tissue regenerating capacity of crosslinked dermal sheep collagen. Journal of biomedical materials research, 28(3), 353–363.
60. Heydarkhan-Hagvall, S., Schenke-Layland, K., Dhanasopon, A. P., Rofail, F., Smith, H., Wu, B. M., Shemin, R., Beygui, R. E., & MacLellan, W. R. (2008). Three-dimensional electrospun ECM-based hybrid scaffolds for cardiovascular tissue engineering. Biomaterials 29(19), 2907–2914.
61. Drury, J. L., & Mooney, D. J. (2003). Hydrogels for tissue engineering: scaffold design variables and applications. Biomaterials 24(24), 4337–4351.

62. Liu, Y., & Chan-Park, M. B. (2009). Hydrogel based on interpenetrating polymer networks of dextran and gelatin for vascular tissue engineering. Biomaterials, 30(2), 196–207.

63. Mosesson, M. W., Siebenlist, K. R., & Meh, D. A. (2001). The structure and biological features of fibrinogen and fibrin. Annals of the New York Academy of Sciences, 936(1), 11–30.

64. Drew, A. F., Liu, H., Davidson, J. M., Daugherty, C. C., & Degen, J. L. (2001). Wound-healing defects in mice lacking fibrinogen. Blood, 97(12), 3691–3698.

65. Sahni, A., & Francis, C. W. (2000). Vascular endothelial growth factor binds to fibrinogen and fibrin and stimulates endothelial cell proliferation. Blood, 96(12), 3772–3778.

66. Koch, S., Flanagan, T. C., Sachweh, J. S., Tanios, F., Schnoering, H., Deichmann, T., Ellä, V., Kellomäki, M., Gronloh, N., Gries, T., Tolba, R., Schmitz-Rode, T., & Jockenhoevel, S. (2010). Fibrin-polylactide-based tissue-engineered vascular graft in the arterial circulation. Biomaterials, 31(17), 4731–4739.

67. Pankajakshan, D., Philipose, L. P., Palakkal, M., Krishnan, K., & Krishnan, L. K. (2008). Development of a fibrin composite-coated poly (ε-caprolactone) scaffold for potential vascular tissue engineering applications. Journal of Biomedical Materials Research Part B: Applied Biomaterials, 87(2), 570–579.

68. Bäckdahl, H., Helenius, G., Bodin, A., Nannmark, U., Johansson, B. R., Risberg, B., & Gatenholm, P. (2006). Mechanical properties of bacterial cellulose and interactions with smooth muscle cells. Biomaterials, 27(9), 2141–2149.

69. Bäckdahl, H., Risberg, B., & Gatenholm, P. (2011). Observations on bacterial cellulose tube formation for application as vascular graft. Materials Science and Engineering: C, 31(1), 14–21.

70. Klemm, D., Schumann, D., Udhardt, U., & Marsch, S. (2001). Bacterial synthesized cellulose artificial blood vessels for microsurgery. Progress in polymer science, 26(9), 1561–1603.

71. Klemm, D., Heublein, B., Fink, H. P., & Bohn, A. (2005). Cellulose: fascinating biopolymer and sustainable raw material. Angewandte Chemie International Edition, 44(22), 3358–3393.

72. Schumann, D., Wippermann, J., Klemm, D., Kramer, F., Koth, D., Kosmehl, H., Wahlers, T., & Salehi-Gelani, S. (2009). Artificial vascular implants from bacterial cellulose preliminary results of small arterial substitutes. Cellulose, 16(5), 877–885.

73. Wan, Y., Gao, C., Han, M., Liang, H., Ren, K., Wang, Y., & Luo, H. (2011). Preparation and characterization of bacterial cellulose/heparin hybrid nanofiber for potential vascular tissue engineering scaffolds. Polymers for Advanced Technologies, 22(12), 2643–2648.

74. (a) Linhardt, R. J., Murugesan, S., & Xie, J. (2008). Immobilization of heparin approaches and applications. Current Topics in Medicinal Chemistry, 8(2), 80–100.
(b) Oliveira, G., Carvalho, L., & Silva, M. (2003). Properties of carbodiimide treated heparin. Biomaterials, 24(26), 4777–4783.

75. Vepari, C., & Kaplan, D. L. (2007). Silk as a biomaterial. Progress in polymer science, 32(8), 991–1007.

76. Bini, E., Knight, D. P., & Kaplan, D. L. (2004). Mapping domain structures in silks from insects and spiders related to protein assembly. Journal of Molecular Biology, 335(1), 27–40.

77. Lovett, M., Eng, G., Kluge, J., Cannizzaro, C., Vunjak-Novakovic, G., & Kaplan, D. L. (2010). Tubular silk scaffolds for small diameter vascular grafts. Organogenesis, 6(4), 217–224.
78. Marelli, B., Alessandrino, A., Fare, S., Freddi, G., Mantovani, D., & Tanzi, M. C. (2010). Compliant electrospun silk fibroin tubes for small vessel bypass grafting. Acta biomaterialia, 6(10), 4019–26.
79. Zhang, X., Baughman, C. B., & Kaplan, D. L. (2008). In vitro evaluation of electrospun silk fibroin scaffolds for vascular cell growth. Biomaterials, 29(14), 2217–2227.
80. Liu, H., Li, X., Niu, X., Zhou, G., Li, P., & Fan, Y. (2011). Improved Hemocompatibility and Endothelialization of Vascular Grafts by Covalent Immobilization of Sulfated Silk Fibroin on Poly (lactic-coglycolic acid) Scaffolds. Biomacromolecules, 12(8), 2914–2924.
81. Liu, H., Li, X., Zhou, G., Fan, H., & Fan, Y. (2011). Electrospun sulfated silk fibroin nanofibrous scaffolds for vascular tissue engineering. Biomaterials, 32(15), 3784–3793.
82. McClure, M., Sell, S., Ayres, C., Simpson, D., & Bowlin, G. (2009). Electrospinning-aligned and random polydioxanone–polycaprolactone–silk fibroin-blended scaffolds: geometry for a vascular matrix. Biomedical Materials, 4(5), 055010.
83. Kidoaki, S., Kwon, I. K., & Matsuda, T. (2005). Mesoscopic spatial designs of nano- and microfiber meshes for tissue-engineering matrix and scaffold based on newly devised multilayering and mixing electrospinning techniques. Biomaterials, 26(1), 37–46.
84. Williamson, M. R., Black, R., & Kielty, C. (2006). PCL–PU composite vascular scaffold production for vascular tissue engineering: attachment, proliferation and bioactivity of human vascular endothelial cells. Biomaterials, 27(19), 3608–3616.
85. In Jeong, S., Kim, S. Y., Cho, S. K., Chong, M. S., Kim, K. S., Kim, H., Lee, S. B., & Lee, Y. M. (2007). Tissue-engineered vascular grafts composed of marine collagen and PLGA fibers using pulsatile perfusion bioreactors. Biomaterials, 28(6), 1115–1122.
86. (a) Zhang, X., Thomas, V., & Vohra, Y. K. (2009). In vitro biodegradation of designed tubular scaffolds of electrospun protein/polyglyconate blend fibers. Journal of biomedical materials research. Part B, Applied biomaterials, 89(1), 135–47.
(b) Zhang, X., Thomas, V., Xu, Y., Bellis, S. L., & Vohra, Y. K. (2010). An in vitro regenerated functional human endothelium on a nanofibrous electrospun scaffold. Biomaterials, 31(15), 4376–81.
87. McClure, M. J., Simpson, D. G., & Bowlin, G. L. (2012). Tri-layered vascular grafts composed of polycaprolactone, elastin, collagen, and silk: Optimization of graft properties. Journal of the Mechanical Behavior of Biomedical Materials, 10, 48–61.
88. Wang, X., Lin, P., Yao, Q., & Chen, C. (2007). Development of small-diameter vascular grafts. World journal of surgery 31(4), 682–689.
89. L'Heureux, N., Dusserre, N., Marini, A., Garrido, S., de la Fuente, L., & McAllister, T. (2007). Technology insight: the evolution of tissue-engineered vascular grafts from research to clinical practice. Nature clinical practice. Cardiovascular medicine 4(7), 389–95.
90. (a) Quiñones-Baldrich, W. J., Prego, A., Ucelay-Gomez, R., Vescera, C. L., & Moore, W. S. (1991). Failure of PTFE infrainguinal revascularization: patterns, management alternatives, and outcome. Annals of Vascular Surgery 5(2), 163–169.

(b) Okoshi, T., Soldani, G., Goddard, M., & Galletti, P. (1993). Very small-diameter polyurethane vascular prostheses with rapid endothelialization for coronary artery bypass grafting. The Journal of Thoracic and Cardiovascular Surgery 105(5), 791.

(c) Seeger, J. M. (2000). Management of patients with prosthetic vascular graft infection. American Surgeon, 66(2), 166–177.

(d) Sarkar, S., Salacinski, H., Hamilton, G., & Seifalian, A. (2006). The mechanical properties of infrainguinal vascular bypass grafts: their role in influencing patency. European journal of vascular and endovascular surgery, 31(6), 627–636.

91. Cummings, C. L., Gawlitta, D., Nerem, R. M., & Stegemann, J. P. (2004). Properties of engineered vascular constructs made from collagen, fibrin, and collagen fibrin mixtures. Biomaterials, 25(17), 3699–3706.

92. McKenna, K. A., Hinds, M. T., Sarao, R. C., Wu, P. C., Maslen, C. L., Glanville, R. W., Babcock, D., & Gregory, K. W. (2012). Mechanical property characterization of electrospun recombinant human tropoelastin for vascular graft biomaterials. Acta biomaterialia, 8(1), 225–233.

93. Lee, J., Yoo, J. J., Atala, A., & Lee, S. J. (2012). The effect of controlled release of PDGF-BB from heparin-conjugated electrospun PCL/gelatin scaffolds on cellular bioactivity and infiltration. Biomaterials, 33(28), 6709–6720.

CHAPTER 3

POLYMERS FOR USE IN THE MONITORING AND TREATMENT OF WATERBORNE PROTOZOA

HELEN BRIDLE and MOUSHUMI GHOSH

CONTENTS

Abstract 50

3.1   Introduction 50

3.2   Waterborne Protozoa 52

3.3   Interactions between (OO)Cysts and Surfaces 55

3.4   Existing Methods of Monitoring and Treatment 56

3.5   Sources of Polymers 59

3.6   Understanding Polymer Protozoa Interactions 61

3.7   Conclusions and Future Outlook 66

Keywords 67

References 67

ABSTRACT

Protozoal pathogens especially *Cryptosporidium* and *Giardia* spp. are recognized worldwide as important pathogens, which cause diarrheal disease in humans and animals. The environmentally stable cysts or oocytes are impervious to inactivation by many drinking water disinfectants and the existing detection techniques are time consuming with low recovery rates. These factors have lead to calls for reliable, rapid and economical approaches for concentrating and simultaneously identifying the (oo) cysts. Use of synthetic organic polymers is an important aspect of water treatment, and bipolymers of biological origin are currently deemed as potential alternatives. Polymeric materials diverse and can concentrate biological entities such as viruses and bacteria and thus can be envisaged as potential candidates for targeting oocytes. Outcomes of systematic studies targeted in elucidating the role of functional groups, wettability, and roughness of cysts in polymer adherence may offer leads for future applications of polymers. In this paper we summarize the existing knowledge on factors governing adherence of oocytes to solid surfaces and polymers. Finally, this chapter makes recommendations of future work bringing together different perspectives of the use of natural and synthetic polymers in different applications.

3.1 INTRODUCTION

Contamination of water by protozoa, such as *Cryptosporidium* and *Giardia* is a serious global issue. These pathogens are ubiquitous in the environment, resistant to standard chlorination disinfection procedures and have a low infectious dose [5]. Furthermore, the presence of bacteria associated with fecal pollution does not indicate the presence of *Cryptosporidium* or *Giardia* spp. in waters. For this reason, regulatory monitoring is frequently undertaken. However, the existing protocols have low recovery rates (13% is considered acceptable) and do not provide information on species of viability of the detected pathogens [16].

Ingestion of human pathogenic species of *Cryptosporidium* oocytes causes cryptosporidiosis, for which there is no safe and effective treatment, and ingestion of *Giardia* cysts causes giardiosis. In developing countries, it is estimated that 250–500 million cryptosporidiosis cases occur each year, playing a significant role in high childhood mortality and morbidity [55]. Prevalence of giardiosis is around 20–30% in the developing world, with up to 100% of children acquiring the infection before the age of 3 [46]. In the developed world, where water treatment is better and more wide-spread, the prevalence is lower but outbreaks do occur. For *Cryptosporidium* one of the most serious outbreaks was in Milawaukee in 1993, and there were several recent cases in the UK, Australia and in Sweden [5]. In the US *Giardia* was the most common intestinal protozoan infection in the early 2000s with infections reported in Norway in 2004.

Understanding the behavior and fate of protozoa in water treatment systems is essential to assess risk at existing plants and appropriately design future systems. Although it is known that the nature of the coagulation pretreatment is very important for the efficiency of the subsequent water treatment processes, the exact adhesion and removal mechanisms have not been elucidated. Few field studies of protozoa in water treatment systems have been undertaken, due to limitations in assay techniques for determining a mass balance for (oo)cysts and lack of understanding of the mechanisms of interaction with chemicals or surfaces within the process. Instead, laboratory studies have concentrated on the adhesion characteristics, to a range of materials, and measurement of interaction forces.

While various studies of *Cryptosporidium* adhesion have been undertaken, with materials ranging from metal oxides, quartz, silanes, natural organic matter, biofilms, clays and natural suspended sediments, little work, apart from a paper by Dai et al who have investigated polymeric materials [13]. Additionally, the majority of studies investigating *Giardia* interactions with surfaces have focused on the postingestion trophozoite stage and its attachment through an adhesive disk. There has been limited investigation of the cyst stage, where the adhesive disk is internalized and fragmented, apart from that reported by Dai [13].

There are several reasons why understanding the interactions of these protozoa with polymers is potentially very useful. Firstly, the membranes employed in filtration methods of water treatment, and monitoring, are made out of polymeric materials; secondly, polymers could easily be used as coatings in water treatment systems or sensing applications; thirdly, it is easy to systematically vary polymer properties to facilitate studies to elucidate structure-activity relationships [6].

In this chapter, we review our latest work investigating the interaction of the protozoa, *Cryptosporidium* and *Giardia*, with a wide range of polymers, of both natural and synthetic origin. This chapter summarizes how comparison of these studies have added to our understanding of which factors control adhesion of protozoa to polymers and explores how this improved knowledge could be exploited in water applications.

3.2 WATERBORNE PROTOZOA

Here we give a detailed introduction to two of the most problematic waterborne protozoa. Other waterborne protozoa exist, including cyclospora, entamoeba and toxoplasma but the below two are responsible for the majority of protozoal disease incidence worldwide.

3.2.1 *CRYPTOSPORIDIUM*

Cryptosporidium was recently placed on the WHO's Neglected Diseases Initiative, and is responsible for a significant proportion of childhood mortality in developing countries [55] and several outbreaks of disease, associated with water treatment failure, in developed countries, including the US [36], UK [14], Australia [40] and Sweden [54]. If ingested, this pathogen can cause an acute self-limiting gastroenteritis, cryptosporidiosis, in immuno-competent hosts and potentially fatal protracted disease in immuno-compromised ones. Research into this pathogen intensified in the 1980 s after its association as a major opportunistic pathogen in patients with AIDS [57]. There is no recognized safe and effective treatment for human cryptosporidiosis [51].

In the developing world, persistent diarrhea, caused by agents such as *Cryptosporidium*, accounts for 30–50% of mortality for children under the age of 5 and it is estimated that 250–500 million cases of cryptosporidiosis occur each year [55]. In the developed world, cryptosporidiosis presents a high risk mainly to the very young, the elderly and immuno-compromised individuals, and accounts for most gastro-intestinal disease outbreaks, where water supplies are chlorinated. The unreported rate of disease from *Cryptosporidium* in England alone has been estimated at >60,000 cases per year [3] with tap water being most common risk factor in recorded cases. *Cryptosporidium* is particularly problematic to the water industry since it is resistant to both environmental stress and standard chlorination disinfection procedures and can survive for up to 16 months in water [10].

Additionally, these organisms are ubiquitous in the environment and have an extremely low infectious dose. For some *C. parvum* isolates, one of the human pathogenic species, less than ten oocytes can be required to cause infection [42, 32]. This number should be compared against the billions of oocytes that an infected host could shed during an episode of infection [52]. During a clinical infection a calf may shed around 10,000 million oocytes, which would provide enough parasites to infect the whole human population of Europe.

In addition to the risk of disease, *Cryptosporidium* has a significant economic impact. For example, a *Cryptosporidium* contamination in the water supply for Sydney, Australia cost US$45 million in direct emergency measures [2], despite no recorded increase in the cryptosporidiosis case rate. Medical expenses and the cost of lost productivity for the Milwaukee outbreak, the largest documented outbreak with over 400,000 people infected, were estimated at US$96 million [12]. There are also substantial economic costs involved in upgrading water treatment plants to deal with the issue of *Cryptosporidium*.

Not all of the >20 *Cryptosporidium* species, and more than 44 genotypes, are pathogenic to humans [48]. Given this it is clear that species identification is an essential characteristic to assess the public health risk arising from detection of any oocytes in a sample. There are no antibodies currently available that can distinguish species differences on the (oo) cysts wall surface [42] and thus genetic comparisons using molecular techniques become important. *C. hominis* and C. *parvum* are the most commonly detected in human clinical cases [52], though several others have been shown to infect humans. *C. hominis* and C. *parvum* have dimensions of 4.5×5.5 µm, and contain four sporozoites (Fig. 3.1 also shows the typical life cycle of this pathogen). *C. parvum* is the major zoonotic species, which causes acute neonatal diarrhea in livestock and is a major contributor to environmental contamination with oocytes [52]. The characteristics of different *Cryptosporidium* species, including (oo)cysts size, host preference and infection sites have been reviewed by Smith and Nichols [52] and we recommend both this review and an earlier one by the same author [53] for further information on this pathogen.

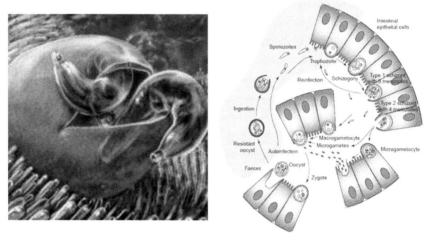

FIGURE 3.1   *Cryptosporidium*. Left: schematic of an (oo)cysts. Right: lifecycle of the pathogen.

3.2.2   GIARDIA

Giardia lamblia (*G. lamblia*) (also known as *G. intenstinalis* or *G. duodenalis*) contaminates water supplies across the globe and ingestion of its cysts can cause giardiasis [8], an acute self-limiting gastroenteritis. This species is the only one in the genus associated with human infection. Like *Cryptosporidium*, *G. lamblia* has a low infectious dose (1–10 cysts) [8] and those most at risk are the young or immune compromised [34]. Treatment of giardiasis varies depending on the patient, as does the effectiveness of different drugs, whose side effects are very common [25]. This pathogen causes a major problem for the water industry as it is resistant to disinfection by chlorine treatment [43] and can also pass with up to 30% efficiency through more advanced membrane filters [17].

Prevalence of *G. lamblia* is around 20–30% in the developing world [18], with up to 100% of children acquiring the infection before the age of 3 [25]. In the developed world, where water treatment is better and more wide-spread, the prevalence is lower but outbreaks do occur. In 1985 there were particularly serious cases in both the UK [28] and US [30]. In the US *G. lamblia* was the most common intestinal protozoan infection in the early 2000s [31] and it has been estimated that there are 2 million cases annually, although many individuals are asymptomatic [8]. More recently, in late 2004, over 1000 cases were reported in Norway, resulting from leaking sewage and ineffective water treatment [41].

G. *lamblia* is zoonotic, with common animal reservoirs being beavers, cattle, cats and dogs [34]. *G. lamblia* cysts are around 8 12 μm in size, containing two trophozoites, and the cyst structure lacks the presence of a noticeable suture line or operculum, which demarcates the site of excystation in other intestinal parasite ova stages [8]. The cyst wall is robust, and protects the internal trophozoites [21]. A review of the details of the cell biology of *Giardia* is given by Gillin and Reiner [22] and studies of the make-up of the cyst wall are still ongoing [8].

3.3 INTERACTIONS BETWEEN (OO)CYSTS AND SURFACES

Various studies of *C. parvum* adhesion have been undertaken, with materials ranging from metal oxides [59], quartz [33, 35], silanes [7], natural organic matter [27], biofilms [50], clays and natural suspended sediments [49] to polymers [13]. The attachment efficiency was generally low, even at high ionic strengths, where Derjaguin, Landau, Verwey and Overbeek (DLVO) theory predicts no energy barrier to adhesion. These results have been explained by the presence of a fluffy glycoprotein layer [35] extending approximately 115 nm from the (oo)cysts wall [7]. This brush-like arrangement of macromolecules contributes "electrosteric" repulsion preventing (oo)cysts adhesion to surfaces. Numerous factors influence the (oo)cysts-surface interaction including (oo)cysts treatment (e.g., formalin or heat treatment to inactivate oocytes or proteinase K to digest surface macromolecules), nature of the surface (charge and hydrophobicity) and solution conditions (e.g., presence of divalent ions or ionic strength).

The principles of colloid and surface science have been exploited to elucidate the interactions of the protozoal cysts to solid surfaces; in particular, sand cyst interactions have proved to be a favorite model since physical removal of protozoa occurs through direct contact with granular medium (filter). Interaction is thought to be a function of surface chemical charge, functionality and cell surface hydrophobicity.

Recently, [58] reported the mechanism of *G. lamblia* (oo)cysts interactions with silica. They described the surface as a polyelectrolyte brush at intermediate separations (5–115 nm from linear compliance) with an electrical double layer often observed at greater separations [58]. The authors proposed the two main repulsive interactions to be electrical and steric in origin. The electrical double layer arises from the presence of charged groups (e.g., ionized organic acids, amines) while the steric interaction originates from surface associated biomolecules (e.g., proteins and/or carbohydrates). Although earlier the electrical double layer was considered within DLVO theory, the DLVO approach alone fails to attain quantitative agreement over the entire range of investigated conditions and this was attributed to factors that include surface roughness, the compressibility of the surface, as well as a steric component arising from the interaction of tethered, long chain biomolecules.

Alternatively, a simple proposition of a protein chain model of the cyst surface qualitatively rationalizes many of the observed interactive effects. The cyst wall protein chain of *G. lamblia* is negatively charged under the conditions of the water treatment plant (pIEP (4) < pH > pKa (−NH2)). Under these conditions the protein chain is elongated and electrosteric attractions are in play. However, when the pH is below the net isoelectric point of the cyst surface, carboxylic groups on acidic amino acid residues are protonated, and there is a slight electropositive charge overall, leading to the collapse of surface protein chain. In fact these findings corroborated the results of Ref. [19] who demonstrated the role of surface carboxylate groups of *C. parvum* oocytes to haematite over a range of solution pH [19].

3.4   EXISTING METHODS OF MONITORING AND TREATMENT

Quantitative microbial risk assessment (QMRA) using *Cryptosporidium* and *Giardia* as representative protozoans (on account of their high prevalence, persistence and resistance to chlorination) has been suggested as an effective component of a multibarrier approach for mitigating waterborne hazards from parasitic protozoans. Through this assessment, variations in source water quality and treatment performance can be evaluated for their contribution to the overall risk; besides assessment of the adequacy of existing control measures, optimization and establishment of appropriate critical control points.

The inactivation of the protozoans in raw water is complicated due to their resistance to commonly used disinfectants such as chlorine. Physical removal methods such as chemically assisted filtration, coagulation, flocculation followed by disinfection are the currently used treatment process.

Advances in, for instance, micro/ultra/nano-filtration and reverse osmosis have been suggested in the treatment process. Demonstration of log reduction credits an challenge and testing results have not been comprehensively documented. Both intrinsic resistance and strain variability of *Cryptosporidium* account for its enhanced tolerance to chlorine over *Giardia*; higher doses of chlorine result in generation of DBPs thus options of using UV, alone or in combination with chlorine/ozone in inactivating the protozoans have been currently reviewed. Appropriate dosing of ozone or UV can be challenging and another problem can be confirmation that the treatment has been effective. (OO)cysts disinfected in this manner appear to be intact when inspected using a microscope; detection methods indicating viability are required.

Regular monitoring using reliable detection techniques in both farmed animals and water is deemed important for its prevention; accurate detection in several cases requires concentration of oocytes prior to direct detection and must be taken into account.

One standard recovery and detection protocol includes a concentration step, usually through filtration, followed by a purification step, involving density gradients, and finally an immunofluorescent staining step which increase the chances that the (oo)cysts can be detected by microscopy. This protocol, "information collection rule" (ICR) for protozoa, was developed as a standardized method for the collection and quantification of *Giardia* cysts and *Cryptosporidium* oocytes [4]. It is still being used by some labs and has been adopted as a standard method for the detection of other parasites as well from effluents, surface and ground water sources. However, due to low recovery rates associated with the ICR method, a newer method called Method 1623 was developed. The main difference is the use of immunomagnetic separation (IMS), which allowed for more efficient separation of the organisms from other debris resulting in a cleaner slide for microscopy [16].

Method USEPA 1623 describes the filtration, concentration and subsequent detection of these pathogens, and illustrated in Fig. 3.2 [16]. Like with fecal indicator organisms, filtration is used to capture the organisms. However, since infectious doses are so low and the organisms will not multiply outside of a host, large water volumes are sampled to increase detection likelihood. Attempts to develop culture methods have not yet met with great success [29]. The first stage thus comprises filtration of 1000 L, over a period of 24 h. This cartridge filter is then eluted to release the (oo) cysts and a second filtration step is performed, this time with a membrane filter. Next the (oo)cysts are subjected to centrifugation, followed by specific separation using immunomagnetic beads. Finally, the (oo)cysts are removed from the beads, placed on a microscope slide and fluorescently stained before they are examined under the microscope by a highly trained technician.

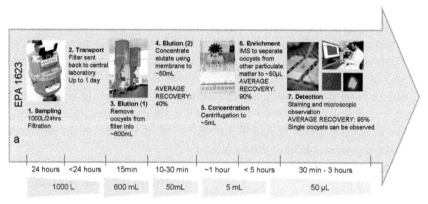

FIGURE 3.2   EPA1623 Method.

The figure illustrates the stages in the EPA1623 method which is widely used for the monitoring of the protozoan (oo)cysts *Cryptosporidium* and *Giardia* (illustration is for *Cryptosporidium*, though the steps are identical), along with the recovery rates and timings.

Each stage of the monitoring process reduces the initial 1000 L to smaller and smaller volumes eventually reaching the 50–100 μL for the microscope slide. Throughout the process there are significant losses of (oo)cysts with acceptable recovery rates for this method at just over 10%. Water companies in the UK report around 30% recovery as an average. It is clear from the above discussion that sample processing is key to the effective monitoring of waterborne protozoan. It is extremely unlikely that any detection technology could reliably detect to the single (oo)cysts level in 1000 L.

Drawbacks of this approach are that time required for detection, the potentially low recovery rates, the need of expensive fluorescent reagents as well as the requirement for highly trained technicians and lab equipment (especially the microscope). The protocol can take up to 3 days and no indication of species or viability is given.

Several emerging techniques for detection have been reported and compared. Some of these have been adopted for instance, a selective dye based method specifically for *Cryptosporidium* spp. oocytes and immuno-based techniques like indirect fluorescent antibody assay or enzyme-linked immunosorbent assay or desoxyribonucleic acid (DNA) detection by polymerase chain reaction (PCR) [4, 20]. Molecular methods have been developed for these pathogens though there is not yet agreement over what are the most appropriate nucleic acid targets, and whether it is possible to determine viability simultaneously. Additionally, reliable single (oo)cysts detection is not yet available.

Many of the detection techniques process μL to a few mL, whereas it may be necessary to sample hundreds of milliliters to thousands of liters. Large sample volumes are required to gain a more representative sample of a large water volume as well as to increase the likelihood of detecting pathogens present at very low concentrations. Sampling of a large volume over a period of time will to some extent compensate for some spatial and temporal variations in pathogen distribution. Additionally, the concentration aspect of sample processing is necessary to bring the pathogen concentration into the detection limit of monitoring methods. Furthermore, while some techniques would detect single pathogens the time required for a single pathogen in a large volume of water to reach the detection area/surface could be prohibitively long [4].

The outcomes emerging from applicability of these techniques however, necessitate a rapid, reliable, high throughput procedure which may simultaneously concentrate oocytes and enable detection and thus may offer realistic solutions for reducing the disease burden of *Cryptosporidium* and *Giardia*.

3.5   SOURCES OF POLYMERS

3.5.1   *BIOPOLYMERS*

Polymeric materials of natural origin have a long history; it is interesting to exploit their association with parasitic protozoans in the natural environment, since the latter also have a history of occurrence in natural

environments. Polymers of biological origin are considered to be 'naturally safe' and degradable and therefore stand out from synthetic polymers
in terms of their consumer acceptability, besides they possess an amazing
structural diversity. An intrinsic ability of such biopolymers is their affinity to a vast number of inorganic, organic natural materials. In fact, natural
polymers (such as chitosan) have been advocated as potential flocculants
as an alternative to organic synthetic polymers for water treatment. Exopolymers of microbial origin are unique in terms of their diverse structure,
robustness, structural plasticity and efficacy.

Microbial exopolymers have remained an area of continued interest
apparently for their diverse applications; consequently a large number of
studies have explored potential microbes for their extracellular biopolymers, which have been characterized and applications suggested. The majority of applications have focused in separation or flocculation of water-
arguably as an alternative to currently used organic synthetic polymers,
which have demonstrable detrimental effect on health.

Few studies have however, attempted to investigate interactions of
such extracellular microbial polymers to protozoal cysts. Recently, Ghosh
et al. (2006) demonstrated excellent flocculating ability of a exopolymer
from *K. terrigena*; the biopolymer effectively flocculated a wide range
of colloid particles (0.5 to 25 μm). The purified biopolymer at low doses,
could remove 62.3% of *Cryptosporidium* oocytes ($1 \quad 10^6$) spiked in tap
water samples over a pH range of 6–8; Calcium (5 mM) was required for
effective removal. Attempts to overproduce the biopolymer through meta-
bolic engineering by the same authors and the possibility of enhancing
selectivity or introducing antimicrobial function by surface modification
of the biopolymer have been explored and were successful (unpublished
observations by Ghosh).

An important question with regards to the feasibility of microbially
produced exopolymers is their low yield. This problem may be circum-
vented by biopolymer overproducing strains. Attempts to overproduce
have been made using Tn5 mutagenesis in putative exopolymer producing strains such as *K.terrigena* and *Acinetobacter*. An over expression of
two intracellular biopolymer synthetic enzymes, phosphoglucomutase and
glucosyl transferase in Tn5 mutants and a 5–7 fold increase in biopolymer
levels was achieved (unpublished observations by Ghosh). The physicochemical and structural signatures of the overproduced purified biopolymer was identical to that produced by the wild type counterpart producer

bacteria. These results are encouraging for exploiting the biopolymer-(oo) cysts interactions extensively. However, specific studies using purified biopolymers for binding/concentrating *Cryptosporidium* or *Giardia* (oo) cysts are yet to be carried out.

3.5.2 SYNTHETIC POLYMER LIBRARIES

Combinatorial chemical synthesis has allowed for development of large polymer libraries containing thousands of different polymers made of up varying combinations, and ratios, of monomers. The interactions of these polymers with a range of biological materials is enabled through a polymer microarray approach, in which small dots of polymer are printed onto microscope slides. Following exposure of the slide to the biological entity of interest, high-throughput imaging technologies can be applied to determine the extent of interactions with each different polymer type.

Polymer microarrays provide a rapid format to enable the screening of a large number of polymers [26, 15]. Such arrays have been applied to determine which materials can enrich, manipulate or modulate a variety of cell types, including human embryonic cells [1, 23, 24], human skeletal progenitor cells [56, 31], human renal tubular epithelial cells and mouse bone marrow dendritic cells [38] suspension cells [44] and bacteria [45].

Bridle was the first to apply the polymer microarray approach to the study of waterborne protozoa, studying both *Cryptosporidium* and *Giardia* [46, 61]. Similarities in adhesion characteristics were identified, with some polymers strongly binding or strongly repelling both protozoa. Further studies are ongoing to determine whether polymers can discriminate between the protozoa, between different species and between viable and nonviable (oo)cysts.

3.6 UNDERSTANDING POLYMER PROTOZOA INTERACTIONS

Unfortunately, little has been researched upon the interactions of polymers and protozoal cysts. The Dai [13] work and recent work by the authors of this chapter remain the exception to this. The inherent structural complexity of the (oo)cysts surface as well as biopolymers/polymers may be a bottleneck in such studies. In one study, water samples containing *C. parvum* oocytes demonstrated high adhesion to a biopolymeric flocculent challenged with a novel biopolymeric flocculent (Ghosh et al, 2006). An explanation to these observations was attributed to both electrostatic and steric forces involved in (oo)cysts-surface association. The binding was thought to occur due to the reduction in steric repulsion mediated by dissolved cations, for instance, Ca^{2+} to the negatively charged (because of the presence of uronic acid) bioflocculant. Binding of cations (Ca^{2+}) to (oo)cysts surface proteins, leading to a neutralization and collapse of such proteins have been previously reported. It is plausible that the presence of calcium-facilitated bioflocculant binding to the oocytes, possibly through entrapment.

However, an exact mechanism through which this exopolysaccharide bioflocculant interact with *Cryptosporidium* (oo)cysts is not yet understood. Clearer understanding of the structure of (oo)cysts wall may help explain or predict interactive phenomenon. Here we also discuss recent findings related to polymer surface properties, for example, wettability or hydrophobicity, surface roughness and polymer composition, which influence adhesion characteristics. Considering together these factors improves our understanding of (oo)cysts polymer interactions, though further work is needed to fully understand the mechanisms and processes.

3.6.1   CRYPTOSPORIDIUM WALL STRUCTURE

Cryptosporidium oocytes are known to exist in two forms, thin walled and thick walled. The role of these alternative structures remains unclear; however, it is believed that thin-walled oocytes reinitiate the parasite developmental cycle within the same host, while thick-walled oocytes are the major form shed into the environment. It has not been determined if the two (oo)cysts walls have different composition. To date, only one apicomplexan (oo)cysts membrane protein has been isolated, *Cryptosporidium* (oo)cysts wall protein 1 (COWP1). Immuno electron microscopy demonstrated that COWP1 is present in the inner (oo)cysts wall and inside wall-forming bodies of mature macrogametes. COWP1 has a striking cysteine periodicity, with cysteine residues spaced roughly every 10 to 12 amino acids due to tandem arrays of two cysteine-rich domains, designated type I and type II domains. It has been proposed that an extensive disulfide-bonded globular structure or intermolecular disulfide bonds provide rigidity to the (oo)cysts wall.

The repeats in COWP1 are characteristic of the extracellular domains of some cell surface adhesive proteins and receptor-like protein kinases, as well as secreted proteins of the extracellular matrix. A possible inter-action of COWP1 with specific functional groups of polymers may be an interesting proposition for elucidating adherence of (oo)cysts polymer interactions.

3.6.2 GIARDIA WALL STRUCTURE

The wall consists of a fibrillar extracellular matrix, lined by a double inner membrane and an outer filamentous wall [47]. This outer wall is around 300 nm thick, and is the most important aspect when considering the adhesive abilities of the cysts. The outer wall is composed of around 43% carbohydrates [37], 86% of which is a novel β-(1–3)-N-acetyl-D-galactosamine homopolymer [21]. The novel galactosamine forms curled fibrils, which bind to cyst wall proteins via internal lectin domains. Bind-ing to these proteins compresses the homopolymers into the narrow, mesh-like structure in fully formed cyst walls [9]. Additionally, the cyst wall of *Giardia* comprises of Leucine rich repeats (LRRs), which are a func-tionally diverse group of proteins related by the ability to participate in protein-protein interactions. LRRs are believed to confer conformational flexibility upon proteins in which they reside, thereby promoting protein-protein interactions.

A clearer picture from the recent findings may be fashioned as a model, which supposes that the cyst wall protein chain of *G. lamblia* is negatively charged under the conditions of the water treatment plant (pIEP (4) < pH > pKa (–NH2)). Under the latter conditions the protein chain is elongated and electrosteric interactions are in play. However, when the pH is below the net isoelectric point of the cyst surface, carboxylic groups on acidic amino acid residues are protonated, and there is a slight electropositive charge overall, leading to the collapse of surface protein chain.

3.6.3 WETTABILITY

Adhesion to solid surfaces of both *C. parvum* and *G. lamblia* has been shown to differ presumably due to their intrinsic surface characteristics. Hydrophilicity was shown to be an important factor for *C. parvum* oocytes whereas surface charge as well as hydrophobicity was suggested crucial for adhesion of *G. lamblia* oocytes to polymer coated glass spheres [13].

3.6.4 SURFACE ROUGHNESS

For bacterial attachment it is known that irregularities that conform to the size of the bacteria increase the adhesion due to maximizing bacteria-surface contact area [2, 60]. If this hypothesis was correct for *G. lamblia*, it would imply that the surface roughness of cysts is likely to be on the order on 1–10 nm (the RMS value above which no adhesion was observed for *G. lamblia*). This agrees with the result reported by Virtanen and colleagues [58]. However, AFM measurements of *G. lamblia* cysts carried out by binding *G. lamblia* cysts on PA104 coated surface, gave a surface roughness of 53 nm [46], though perhaps this result should be remeasured in light of the Virtanen data.

3.6.5 POLYMER COMPOSITION/FUNCTIONAL GROUPS

Bridle and co-workers used a polymer microarray approach to analyze the binding of (oo)cysts to a range of synthetic polymers. Some previous studies have indicated that the inhibition of cyst binding was strongest in polyacrylates containing Dimethylacrylamide (DMAA), Diethylacrylamide (DEAA) or styrene, as well as selected polyurethanes. Monomers promoting strong binding were more variable; however, the presence of DMAEA (2-(Dimethylamino) ethyl acrylate), DEAEA (2-(Diethylamino) ethyl acrylate), DMAEMA (2-(Dimethylamino) ethyl methacrylate) and DEAEMA (2-(Diethylamino) ethyl methacrylate) was very common among the polymers strongly binding to (oo)cysts [46, 61].

For cellular adhesion it has been reported that glycol functionalities act in a preventative manner [11, 62]. This is normally attributed to the protein repellent nature of these moieties; for the majority of cell types adhesion is considered to occur via initial protein adsorption, which subsequently mediates cellular adhesion. For the protozoan experiments reported here prior protein interaction with the surface is not thought to be a likely mechanism of adhesion given that the experiments are performed in water and the (oo) cysts are not thought to secrete proteins (recent work by Tyler may contradict this for cysts, personal communication) However, the repellent nature of glycol functionalities is still consistent with our results, since none of the polyurethanes, containing monomers with glycols, exhibited strong interactions with cysts. In this case, the known poor likelihood of protein interaction with glycol moieties could apply to the cyst surface proteins, thus limiting any interactions between these polymers and the cyst outer wall.

Aromatic functionalities were correlated with low cell adhesion whereas amine and ester moieties were found to promote cellular adhesion [62] the monomers DMAEMA, DMAEA, DEAEMA or DEAEA, present in the 'hit' array in polymers also containing MEMA and MMA, all contain secondary amine groups and are associated with high levels of cyst adhesion. Previous work has also shown that these monomers promote leucobinding [39]. For (oo)cysts adhesion, the hypothesis is that at physiological pH values, the amines will be protonated and thus ion-pair with the cyst wall. DMAA and DEAA contain amide groups and are present in polymers, which prevent adhesion. Since amide groups will not protonated at physiologically relevant pH this explains the lack of interaction. The DLVO theory considers the balance between two major forces controlling adhesion in colloidal systems; the short-range attractive Lifshitz-van der Waals forces and the electrostatic double-layer forces. The zeta potential of small polymer dots on the microarray is difficult to determine; however, we predict that the polymers utilized in this study will be negatively charged at the experimental conditions, with little variation in zeta potential. (oo)cysts have an isoelectric of around pH2–3 and are therefore negatively charged in the environment, as in the majority of our experiments. Given that both the (oo)cysts and the polymer surfaces have a negative charge a considerable energy barrier to adhesion is predicted at low ionic strengths. Therefore, one would expect that at low pH and high ionic strengths a greater interaction between cysts and the polymers would be observed. Additionally, little variation between polymers is predicted since all zeta potentials are estimated to lie within a narrow range. Neither of the above hypotheses is proven correct from the polymer microarray results. Considerable variation in adhesion characteristics is observed and experiments performed at pH2 reveal that the interaction of *G. lamblia* cysts with polymer surfaces is reduced at this pH.

For the waterborne protozoan, *C. parvum*, which has been more extensively studied, deviation from the DLVO theory predicted behavior has been observed with quartz substrates. Researchers have attributed this to the repulsion between glycoproteins, extending from the (oo)cysts surface that would occur should this layer be compressed to enable the (oo)cysts to closely approach the surface. This steric repulsion explains the absence of (oo)cysts adhesion at high monovalent ionic strengths. Divalent cations are capable of improved compression of the glycoprotein layer associated with improved adhesion characteristics. Additionally, cleaving of these surface groups, using for example the aggressive enzyme, proteinase K, which breaks peptide bonds adjacent to amino acids where the α-amino groups are blocked, and over time degrades proteins to free amino acids, results in greater surface interaction. *G. lamblia* cyst walls are considered more mesh-like structures and therefore a variation in the observed effects may be anticipated.

For polysaccharide polymers the binding may occur due to the reduction in steric repulsion mediated by dissolved cations, for instance, Ca^{2+} to the negatively charged (because of the presence of uronic acid) biopolymer. Binding of cations (Ca^{2+}) to (oo)cysts surface proteins, leading to a neutralization and collapse of such proteins have been reported earlier and may account for the observed adhesion of oocytes to biopolymer surfaces.

3.7   CONCLUSIONS AND FUTURE OUTLOOK

In conclusion we find that while polymers have many potential applications in both water treatment and monitoring the interactions of these materials with waterborne protozoan pathogens has been little studied. Understanding the characteristics controlling adhesion and release of these protozoa to polymers is essential for the appropriate design of water treatment and monitoring technologies.

Recent work by the authors of this chapter has shed new light on the interactions of both *Cryptosporidium* and *Giardia* with a wide range of polymer materials. Work by Ghosh has focused on the use of biopolymers, advantageous for their natural origin and possibility of green synthesis methods, and the application of these in flocculation and treatment. Work by Bridle has employed synthetic polymers in a microarray approach for the simultaneous screening of hundreds of polymers, allowing for the evaluation of characteristics such as wettability, surface roughness and polymer composition as factors controlling adhesion.

Polymers may prove to be a more versatile and novel option for detection and subsequent monitoring of waterborne protozoans. Based on the structural and mechanical properties for polymers, chemical modifications may further ensure selectivity and improve adhesion of oocytes. For biopolymers, with a polysaccharide backbone, engineering of the functional groups offers the possibility to generate new or modified polysaccharide variants with ability to both to concentrate and detect oocytes from diverse environments. Alternately, approaches such as polymer engineering, novel formulation designing, or by linking this polysaccharide with other synthesized polymers may offer viable solutions.

Further work is required to deepen our understanding of how to best design polymers for these applications. Potential ways forward include the use of microarray screening for biopolymers, further microarray studies concentrating in identifying the ability of polymers to discriminate based on differences in species and viability and in evaluating the performance of identified polymers in water monitoring and treatment applications.

KEYWORDS

- **Cryptosporidium**
- **Detection**
- **Giardia**
- **Polymers**
- **Water protozoa**
- **Water quality monitoring**
- **Waterborne pathogens**

REFERENCES

1. Anderson, D. G., Levenberg, S. et al. (2004). "Nanoliter-scale Synthesis of Arrayed Biomaterials and Application to Human Embryonic Stem Cells." Nat Biotechnol, 22(7), 863–866.
2. Bazaka, K., Jacob, M. V. et al. (2011). "Plasma-assisted Surface Modification of Organic Biopolymers to Prevent Bacterial Attachment." Acta Biomaterials, 7, 2015–2028.

3. Bridge, J. W., Oliver, D. M. et al. (2010). "Engaging with the Water Sector for Public Health Benefits Waterborne Pathogens and Diseases in Developed Countries," Bulletin of the World Health Organization, 88, 873–875.

4. Bridle, H. (2013). Waterborne Pathogens: Detection Methods and Applications, Elsevier.

5. Bridle, H. M., Kersaudy-Kerhoas, W., et al. (2012). "Detection of Cryptosporidium in Miniaturised Fluidic Devices," Water Research, 46(6), 1641–1661.

6. Bridle, H., Wu, M. et al. (2011). "Targeting Cryptosporidium Capture" Water Research Accepted.

7. Byrd, T. L., & Walz, J. Y. (2007). "Investigation of the Interaction Force between Cryptosporidium parvum (oo)cysts and Solid Surfaces" Langmuir, 23, 7475–7483.

8. Caccio, S. M., Thompson, R. C. et al. (2005). "Unraveling Cryptosporidium and Giardia epidemiology." Trends Parasitol, 21(9), 430–437.

9. Chatterjee, A., A., Carpentieri, et al. (2010). "Giardia Cyst Wall Protein 1 is a Lectin that Binds to Curled Fibrils of the GalNAc homopolymer." PLoS Pathogens, 6(8).

10. Chen, F., Hunag, K. et al. (2007). "Comparison of viability and infectivity of Cryptosporidium parvum (oo)cysts stored in potassium dichromate solution and chlorinated tap water." Veterinary Parasitology, 150(1–2), 13–17.

11. Cima, L. G. (1994). "Polymer Substrates for Controlled Biological Interactions." Journal of Cellular Biochemistry, 56(2), 155–161.

12. Corso, P. S., Kramer, M. H. et al. (2003). "Cost of Illness in the (1993) Waterborne Cryptosporidium Outbreak, Milwaukee, Wisconsin." Emerging Infectious Diseases, 9(4), 426–431.

13. Dai, X., Boll, J. et al. (2004). "Adhesion of Cryptosporidium Parvum and Giardia lamblia to Solid surfaces the role of surface charge and hydrophobicity." Colloids and Surfaces B Bio interfaces, 34, 259–263.

14. Davies, A. P., & Chalmers, R. M. (2009). "Cryptosporidiosis" British Medical Journal, 339, 4168.

15. Diaz-Mochon, J. J., Tourniaire, G. et al. (2007). "Microarray Platforms for Enzymatic and Cell-Based Assays." Chem Soc Rev, 36(3), 449–457.

16. EPA (2005). Method 1623 Cryptosporidium and Giardia in Water by Filtration/IMS/FA, United States Environmental Protection Agency.

17. Falk, C. C., Karanis, P. et al. (1998). "Bench Scale Experiments for the Evaluation of a Membrane Filtration Method for the Recovery Efficiency of Giardia and Cryptosporidium from Water." Water Research, 32(3), 565–568.

18. Farthing, M. J. G. (1994). Giardiasis as a Disease, Giardia: From molecules to disease. Thompson, R. C. A., Reynoldson, J. A. & Lymbery, A. J., Wallingford, CABI Publishing: 15–37.

19. Gao, X., & Chorover, J. (2009). "In Situ Monitoring of Cryptosporidium Parvum (oo) cysts Surface Adhesion Using ATR–FTIR spectroscopy." Colloids and Surfaces B: Biointerfaces, 71(2), 169–176.

20. Gavriilidou, D., & Bridle, H. (2012). "Comparison of immobilization strategies for Cryptosporidium parvum immunosensors, "Biochemical Engineering Journal, 68, 231–235.

21. Gerwig, G. J., van Kuik, A. et al. (2002). "The Giardia intestinalis filamentous cyst wall contains a novel beta (1–3)-N-acetyl-D-galactosamine polymer: a structural and conformational study." Glycobiology, 12(8), 499–505.
22. Gillin, F. D., & Reiner, D. S. (1996). "Cell Biology of the Primitive Eukaryote Giardia Lamblia." Annual Reviews of Microbiology, 50, 679–705.
23. Hay, D. C., Pernagallo, S. et al. (2009). A Simple Polyurethane Matrix Promotes Hepatic Endoderm Viability and Inducible Drug Metabolism: Implications for Drug Toxicology Testing and the Design of Liver Support Devices. Hepatology.
24. Hay, D. C., Pernagallo, S. et al. (2011). "Unbiased screening of polymer libraries to define novel substrates for functional hepatocytes with inducible drug metabolism." Stem Cell Research, 6(2), 92–102.
25. Heresi, G. P., Murphy, J. R. et al. (2000). "Giardiasis" Seminars in Pediatric Infectious Diseases, 11(3), 189–195.
26. How, S. E., Yingyongnarongkul, B. et al. (2004). "Polyplexes and Lipoplexes for Mammalian Gene Delivery: From Traditional to Microarray Screening." Combinatorial Chemistry and High Throughput Screening 7, 423–430.
27. Janjaroen, D., Liu, Y. et al. (2010). "Role of Divalent Cations on Deposition of Cryptosporidium parvum (oo)cysts on Natural Organic Matter Surfaces" Environ Sci Technol, 44, 4519–4524.
28. Jephcott, A. E., Begg, N. T. et al. (1986). "Outbreak of giardiasis associated with mains water in the United Kingdom" Lancet, 29, 1(8483), 730–732.
29. Karanis, P., & Aldeyarbi, H. M. (2011). "Evolution of Cryptosporidium in Vitro Culture" International Journal for Parasitology 41(12), 1231–1242.
30. Kent, G. P., & Greenspan, J. R. et al. (1988). "Epidemic Giardiasis Caused by a Contaminated Public Water Supply" American Journal of Public Health, 78(2), 139–143.
31. Khan, F., Tare, R. S., et al. (2010). "Strategies for Cell Manipulation and Skeletal Tissue Engineering using High-Through Put Polymer Blend Formulation and microarray techniques." Biomaterials, 31(8), 2216–2228.
32. King, B. J., & Monis, P. T. (2007). "Critical Processes Affecting Cryptosporidium (oo) cysts Survival in the Environment." Parasitology, 134, 309–323.
33. Kuznar, Z. A., & Elimelech, M. (2004). "Adhesion Kinetics of Viable Cryptosporidium parvum (oo)cysts to Quartz Surfaces" Environ. Sci. Technol, 38, 6839–6845.
34. Lebwohl, B., Deckelbaum, R. J. et al. (2003). "Giardiasis" Gastrointestinal Endoscopy, 57(7) 906–913.
35. Liu, Y., Kuhlenschmidt, M. S. et al. (2010). "Composition and Conformation of Cryptosporidium parvum (oo)cysts Wall Surface Macromolecules and their Effect on Adhesion Kinetics of (oo)cysts on Quartz Surface" Bio macromolecules, 11, 2109–2115.
36. MacKenzie, W. R., NHoxie, N. J. et al. (1994). "Cryptosporidium Infection from Milwaukee's Public Water Supply" New England Journal of Medicine, 331(3), 161–168.
37. Manning, P., Erlandsen, S. L. et al. (1992). "Carbohydrate and Amino Acid Analyzes of Giardia muris cysts." J. Protozool, 39(2), 290–296.
38. Mant, A., Tourniaire, G. et al. (2006). "Polymer microarrays Identification of substrates for phagocytosis assays." Biomaterials, 27(30), 5299–5306.
39. Mizomoto, H. (2004). The Synthesis and Screening of Polymer Libraries Using a High Throughput Approach, School of Chemistry. Southampton, University of Southampton. PhD.

40. Ng, J. S. Y., Pingault, N. et al. (2010). "Molecular characterization of Cryptosporidium outbreaks in Western and South Australia." Experimental Parasitology, 125(4), 325–328.

41. Nygard, K., Schimmer, B. et al. (2006). "A large community outbreak of waterborne giardiasis-delayed detection in a nonendemic urban area." BMC Public Health, 6, 141.

42. Okhuyzen, P. C., Chappell, C. L. et al. (1999). "Virulence of three distinct Cryptosporidium parvum isolates for healthy adults." The Journal Of Infectious Diseases, 180(4), 1275–1281.

43. Owen, R. L. (1980). "The ultrastructural basis of Giardia function." Trans R Soc Trop Med Hyg, 74(4), 429–433.

44. Pernagallo, S., Unciti-Broceta, A. et al. (2008). "Deciphering Cellular Morphology and Biocompatibility using Polymer Microarrays." Biomed Mater, 3(3), 034112, 6pp.

45. Pernagallo, S., Wu, M. et al. (2011). "Colonizing new frontiers-microarrays reveal biofilm modulating polymers." J Mater Chem, 21(1), 96–101.

46. Pickering, H., Wu, M. et al. (2012). "Analysis of Giardia lamblia Interactions with Polymer Surfaces using a Microarray Approach." Environ Sci Technol 46(4), 2179–2186.

47. Reiner, D. S. (2008). Cell Cycle and Differentiation in Giardia Lamblia. Department of Microbiology, Tumor and Cell Biology. Stockholm, Karolinska Institutet. PhD.

48. Robinson G., Elwin K., et al. (2008). "Unusual Cryptosporidium genotypes in human cases of diarrhea." Emerging Infectious Diseases, 11, 1800–1802.

49. Searcy, K. E., Packman, A. I. et al. (2005). "Association of Cryptosporidium parvum with Suspended Particles: Impact on (oo)cysts Sedimentation." Applied and Environmental Microbiology 71(2), 1072–1078.

50. Searcy, K. E., Packman, A. I. et al. (2006). "Capture and Retention of Cryptosporidium parvum (oo)cysts by Pseudomonas aeruginosa Biofilms." Applied and Environmental Microbiology, 72(9), 6242–6247.

51. Smith, H. V., & Nichols, R. A. B. (2009). "Cryptosporidium: Detection in water and food." Experimental Parasitology 124(1), 61–79.

52. Smith, H. V., &. Nichols, R. A. B. (2010). "Cryptosporidium: Detection in water and food." Experimental Parasitology 124(1), 61–79.

53. Smith, M., & Thompson, K. C. (2001). Cryptosporidium the Analytical Challenge, Royal Society of Chemistry.

54. Smittsky dds Institutet. (2010). "http: //www.smittskyddsinstitutet.se/sjukdomar/ cryptosporidium-infektion." from http://www.smittskyddsinstitutet.se/sjukdomar/ cryptosporidium-infektion.

55. Snelling, W. J., Xiao, L. et al. (2007). "Cryptosporidiosis in Developing Countries." J. Infect Dev Ctries 1(3), 242–256.

56. Tare, R. S., Khan, F. et al. (2009). "A microarray approach to the identification of polyurethanes for the isolation of human skeletal progenitor cells and augmentation of skeletal cell growth." Biomaterials 30(6), 1045–1055.

57. Tzipori, S., & Widmer, G. (2008). "A hundred-year retrospective on cryptosporidiosis" Trends in Parasitology 4, 184–189.

58. Virtanen, A., Considine, R. et al. (2012). "Direct force measurement between biocolloidal Giardia lamblia cysts and colloidal silicate glass particles." Langmuir 28(49), 17026–17035.

59. Walker, M. J. (1999). "Sorption of Cryptosporidium Parvum (oo)cysts in Aqueous Solution to Metal Oxide and Hydrophobic Substrates" Environ Sci. Technol. 33, 3134–3139.

60. Whitehead, K. A., Colligon, J. et al. (2005). "Retention of microbial cells in substratum surface features of micrometer and submicrometer dimensions." Colloids and Surfaces B: Bio interfaces 41(2–3), 129–138.

61. Wu, M., Bridle, H. et al. (2012). "Targeting Cryptosporidium capture" Water Research 46(6), 1715–1722.

62. Yang, J., Mei, Y. et al. (2010). "Polymer surface functionalities that control human embryoid body cell adhesion revealed by high throuput surface characterization of combinatorial material microarrays" Biomaterials 31(34), 8827–8838.

CHAPTER 4

SYNTHESIS OF POLYPYRROLE/ TIO₂ NANOPARTICLES IN WATER BY CHEMICAL OXIDATIVE POLYMERIZATION

YANG TAN, MICHEL B. JOHNSON, and KHASHAYAR GHANDI

CONTENTS

Abstract 74

4.1   Introduction 74

4.2   Experimental 78

4.3   Instruments 79

4.4   Results and Discussion 79

4.5   Conclusions 93

Acknowledgments 94

Keywords 95

References 95

ABSTRACT

Polypyrrole (PPy) based composites and their conjugates to natural polymers are promising materials for applications in photovoltaic devices, photocatalysts, batteries, and supercapacitors. Polypyrrole/TiO$_2$ nanocomposites were prepared by chemical oxidative polymerization. The nanocomposites were characterized with Fourier-transform infrared (FTIR), thermogravimetric analysis (TGA), scanning electron microscope (SEM), transmission electron microscopy (TEM), and other methods. A new method was developed to make composites with a core-shell structure and with TiO$_2$ in the rutile phase for the first time. The thermal stability of the composite was remarkably enhanced compared to PPy. The optical properties of the composite are different with the optical properties of other types of PPy/TiO$_2$ nanocomposites made by chemical polymerization in water: mostly due to the different sizes of the core and shell of TiO$_2$/PPy nanocomposites, UV-vis spectrum shows a shift in the absorption wavelength.

4.1 INTRODUCTION

Lately, there has been significant attention to the production of materials with multiscale porosity [1]. The multiscale porosity in particular has been found to considerably improve the photocatalytic activity of catalysts [1–3]. As part of our efforts to make nanocomposites based on biomaterials, [4–8] we have started to investigate the incorporation of PPy/TiO$_2$ nanocomposites in cellulosic material to make multi scale biodegradable photoactive material. In this work, as a first step we study the synthesis of PPy/TiO$_2$ nanocomposites.

We will first briefly introduce PPy and its synthesis. Then we will review the works that have been done to make PPy on the surface of TiO$_2$ nanoparticles (PPy/TiO$_2$ nanocomposites) by chemical methods in water. Then we will discuss our methods and our results and compare those with the existing data on PPy/TiO$_2$ nanoparticle composites made by chemical methods in water.

PPy is a conjugated polymer where delocalized π-electron orbitals overlap along the polymer chains. Doping can involve the addition of electrons or the removal of electrons from a polymer. The charge carriers (in doped PPy) will move along the polymer chain under the influence of electric field. When the charge carriers reach any defect point or the end of polymer chain, charge transfer between neighbor chains occur by hopping or tunneling [9]. The conductivity of PPy is dominated by the intrachain and interchain mobility of charge carriers.

When they are neutral, conjugated polymers are insulators or semi-conductors. Upon oxidation (p-doping) or reduction (n-doping), interband transitions form between the conduction and the valence band to lower the band gap, forming charge carriers along the polymer backbone. Neutral PPy is categorized as an insulator because its band gap is 3.16 eV (a high value) and almost no electron can "jump" from the valence to the conduction band at room temperature. The proposed electronic energy diagram is shown in Fig. 4.1 [10–12].

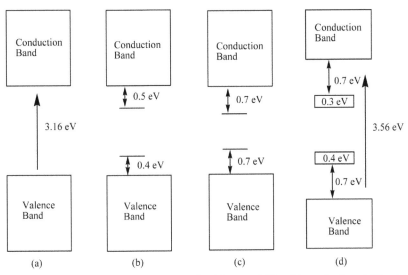

FIGURE 4.1 Electronic energy diagrams for (a) neutral PPy, (b) partly doped PPy, (c) further doped PPy, and (d) fully doped PPy [12].

PPy was first synthesized in 1916 by chemical polymerization with acidified hydrogen peroxide as an oxidant. At the time it was called "pyrrole black" [13]. Due to the cross-linking and other conformational defects of PPy, it is almost insoluble in every solvent tested to date [13]. A summary of some physical properties of PPy is given in Table 4.1 [13].

TABLE 4.1   Some Basic Physical Properties of PPy

Molecular Weight	Unknown due to cross-linking
Crystallinity	Low
Glass Transition Temperature	Variable
Melting Point Temperature	None, decomposes without melting
Appearance	Blue-black solid film (electrochemical) or powder (chemical synthesis)

The major breakthrough regarding the synthesis of PPy was achieved by Diaz and co-workers [14–16] when they reported that highly conductive and stable PPy was prepared under controlled electrochemical conditions [17]. Since then, PPy has been mainly prepared by electrochemical polymerization. However, along with chemical polymerization, other preparation methods for PPy have been reported, involving photochemistry [18], plasma [19], radiolysis [20], and concentrated emulsions [21, 22].

Chemical polymerization of pyrrole is a good method for industry because it offers the advantage of large-scale production. PPy can be prepared easily in various solvents such as aqueous solutions [23], ionic liquids [24], ether [25] and acetonitrile [26]. The physical properties (e.g., morphology, conductivity, environmental stability and optical properties) obtained by chemical oxidative polymerization are strongly dependent on the reaction conditions [27–29].

Nanoscale materials, including those that involve PPy, have been investigated extensively due to their unique properties. Many properties of materials change when their size approaches the nanoscale. One of the most remarkable properties is the large surface-to-volume ratio of nanomaterials, which dramatically improves the surface area and physical properties such as the ability to absorb solar radiation, making them competitive materials as photo catalysts [30]. Other significant size-dependent properties are observed, including optical properties, quantum confinement in semiconductor particles, surface plasmon resonance in some metal particles, and super para magnetism in magnetic materials. Among nanoparticles, especially nano metal oxides, TiO_2 nanoparticles receive a lot of interest because of their chemical stability, low price, nontoxicity [31], and long-term photostability [32].

The commonly known polymorphs of TiO$_2$ found in nature are anatase (tetragonal), brookite (orthorhombic) and rutile (tetragonal) [33]. The structures are shown in Fig. 4.2. Rutile and anatase are the most commonly synthesized phases, but they are thermodynamically and structurally different, resulting in different properties. As an example, the rutile phase is known for its higher chemical stability and refractive index, and lower photo reactivity than anatase [34].

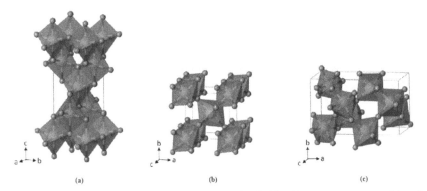

FIGURE 4.2   Structures of three common crystals of TiO$_2$: (a) anatase structure, (b) rutile structure, and (c) brookite structure [35].

Nano composites can acquire novel properties with the proper combination of two or more kinds of nanoparticles or nano objects [36]. Suitable combinations of nano composites can lead to a wide range of potential practical applications. Inorganic nanoparticles might be able to solve organic polymer shortcomings such as inadequate mechanical strength. Therefore, much research has been done to study the polymeric nanocomposites.

Recently, there have been a lot of improvements in electrical conductivity, thermal stability, and processing of organic-inorganic hybrid nanocomposites of PPy metal oxides [37], such as SiO$_2$ [38], ZrO$_2$ [39], Y$_2$O$_3$ [40], and Al$_2$O$_3$ [41]. TiO$_2$ is a very promising material for the synthesis of organic-inorganic hybrid nanocomposites. PPy/TiO$_2$ nanocomposites could obtain advantages from both TiO$_2$ nanoparticles and conducting polymers, leading to various potential applications in photovoltaic cells, light emitting diodes, and photo catalysts [42–44]. For example, TiO$_2$ is a large band gap semiconductor, with band gaps of 3.20, 3.02, and 2.96 eV for the anatase, rutile, and brookite phases, respectively [33]. PPy has narrower band gap compared to TiO_2, thus PPy would be an excellent candidate for binding to TiO_2 to improve the electric or photoelectric properties of composites [8, 45].

In order to make PPy/TiO_2 nano composites, in this work we did the synthesis in ferric chloride aqueous solutions. The synthesis procedure involves: (i) the synthesis of TiO_2 nanoparticles, and (ii) the synthesis of PPy/TiO_2 nanocomposites in deionized water with chemical polymerization.

4.2   EXPERIMENTAL

4.2.1   MATERIALS

Oxidant anhydrous ferric chloride (Sigma-Aldrich), $TiCl_4$ (99.9%, Sigma-Aldrich), nitric acid (68~70%, ACP chemical Inc.) and acetone (99.5%, Caledon Laboratories Ltd.) were used as received. The water that was used for the synthesis was deionized (Purity up to 18.2 MΩ-cm). Pyrrole monomer (98%, Sigma-Aldrich) was purified by reduced-pressure distillation and was kept in the refrigerator prior to use.

4.2.2   SYNTHESIS OF TIO_2 NANOPARTICLES

The synthesis of TiO_2 nanoparticles was done based on a slightly modified literature procedure [46]. 7 mL nitric acid (70%) was added to 120 mL deionized water to adjust the acid concentration to 1 mol L^{-1}. $TiCl_4$ (1 mL) was added slowly to this solution at room temperature. The solution was oven-heated to 95 °C for 24 h. The solid TiO_2 nanoparticles were produced and washed with deionized water and the subsequent suspension was centrifuged at 5000 rpm for 10 min. The XRD of TiO_2 nanoparticles showed that the prepared TiO_2 nanoparticles are in the rutile phase.

4.2.3   SYNTHESIS OF POLYPYRROLE/TIO_2 NANOPARTICLES

The PPy/TiO_2 nanoparticles were prepared in an aqueous medium according to the following method: 0.01 g rutile TiO_2 nanoparticles and 0.3 mL pyrrole were added to 40 mL aqueous FeCl$_3$ (0.7 g) solution. The mole ratio of PPy to TiO$_2$ is 40:1. Polymerization was initiated as soon as the pyrrole was mixed with the FeCl$_3$ solution, and allowed to stir for 8 h at room temperature. The product precipitated from the solution as dark powders. This product was filtered and washed with distilled water. The sample was dried at 70 °C overnight. Approximately 3 mol% TiO$_2$ was included in the PPy/TiO$_2$. The detailed information will be discussed in TGA section.

4.3   INSTRUMENTS

The morphologies of the samples were obtained by scanning electron microscopy (SEM) and transmission electron microscopy (TEM). A JEOL JSM-5600 SEM was used to obtain the images of nano composites. A JEOL 2011 scanning transmission electron microscopy (TEM) was used to obtain TEM micrographs. Elemental analysis of the samples was performed by energy dispersive X-ray spectroscopy.

A Fourier transform infrared spectrometer (FTIR, Nicolet FT-Infrared 200 Spectrometer) was used to characterize the functional groups in the samples over the 4000 to 400 cm^{-1} range at a resolution of 4 cm^{-1}. Thermo gravimetric analysis (TGA, SDT Q600 from TA Instruments) was performed from room temperature to 800°C with a heating rate of 10 °C min^{-1} and an argon flow (50 mL min^{-1}). UV-vis absorption spectra were recorded using a Cary 100 UV-Vis spectrophotometer over the range 200 to 900 nm. Differential scanning calorimetry (DSC, TA Q200 from TA Instruments) thermograms were recorded over the temperature range 20 to 160 °C using a heating and cooling rate of 20 °C min^{-1} under helium flow at a rate of 25 mL min^{-1}.

4.4   RESULTS AND DISCUSSION

4.4.1   MICROSCOPY

4.4.1.1   SEM AND EDS STUDY

The sample's morphology includes many particles and a few films. The morphology of composites depends on factors such as do pants and solvents. For example, use of Fw (β-NAS) as a dopant in the synthesis of

PPy/TiO$_2$ composites, yields nanotube composites [47]. The dissolution of pyrrole in β-NAS solution is an important factor to obtain the tubular morphology due to the hydrotropic behavior of β-NAS as an anionic surfactant [48]. Because this anionic surfactant has amphiphilic nature, pyrrole and even micelles can be dissolved in the solution. When the β-NAS is used in low concentration, granular morphology is preferred. The tubular morphology can only be obtained when the concentration of β-NSA is large enough. We speculate that when there is enough β-NSA in the solution, pyrrole monomers can be entirely dissolved and aggregated orderly around this anionic surfactant, allowing the growth of PPy to form amorphous PPy-NSA nanotubules. When nonhydrotrope acids, such as HCl and p-dodecylbenzenesulfonic acid (DBSA), are used as dopants in the polymerization, granular rather than tubular composites are obtained. This should be the case as well when no acid is used (such as the condition of our reaction). With our synthesis method, pyrrole monomers are polymerized on the TiO$_2$ nanoparticle surfaces, forming core-shell structured nanocomposites (Fig 4.3).

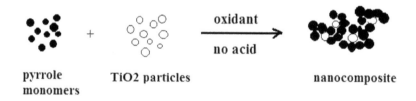

pyrrole monomers   TiO2 particles   nanocomposite

FIGURE 4.3 Schematic diagram of chemical polymerization of pyrrole on the surface of TiO$_2$ nanoparticles.

As shown in the SEM micrograph (Fig. 4.4), sphere-like structures connect with each other in a 3D-network, comprising a porous structure. The EDS image (Fig. 4.5) shows the presence of titanium in the structure. The observed morphology is a result of our synthesis conditions, that is, use of FeCl$_3$ aqueous solution and the stirring during polymerization. Pyrrole and pyrrole oligomers have extremely low solubility in FeCl$_3$ aqueous solution, so pyrrole and its oligomers cannot aggregate with a certain order to form a tube structure. The template for polymerization is around the spheres of TiO$_2$ nanoparticles. Stirring also decreases the chance of an orderly aggregation. The average dimension of the pores is around 1 μm, and the average size of particles is around 300 nm. This porous structure can play an important role in some applications. For example, for photo catalytic and photovoltaic applications the porous structure can increase the surface area.

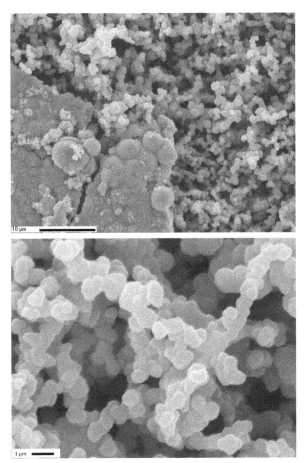

FIGURE 4.4   The SEM micrographs of PPy/TiO$_2$ nanocomposite showing a 3-dimensional network. Micrographs were collected using an accelerating voltage of 15 kV. The samples were coated with gold prior to SEM analysis.

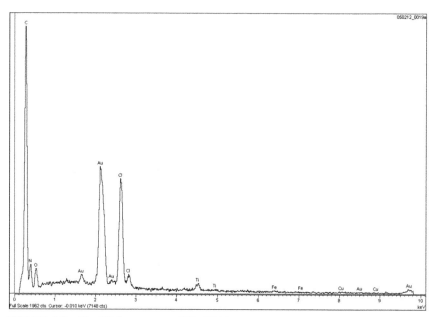

FIGURE 4.5   The EDS spectrum of PPy/TiO2 nanocomposite. The samples were coated with gold prior to characterization.

The composition of the PPy/TiO$_2$ composites influences the morphology as well. Babazadeh et al. [49] reported a method for synthesis of a series of polypyrrole/titanium dioxide (PPy/TiO$_2$) nanocomposites. These nanocomposites were prepared by in situ deposition oxidative polymerization of pyrrole hydrochloride using ferric chloride (FeCl$_3$) as an oxidant in the presence of ultra-fine grade powder of anatase TiO$_2$ nanoparticles cooled in an ice bath. Their synthesis method for PPy is similar to ours, but HCl solution was not used in our research. Avoiding use of caustic acid can make the synthesis process greener. Also for us it is important to use conditions that are nondestructive for biomaterial as our future plans involve incorporation of and PPy/TiO$_2$ in biomaterial.

Figure 4.6 shows the SEM images of PPy and PPy/TiO$_2$ from Babazadeh's work, which is different with ours. In their work, the pure PPy is like a film, and the doping of TiO$_2$ changes PPy's morphology and with the increase of TiO$_2$ contents, the composites transform from film to particle morphology. Figure 4.6c is the SEM image of their composite with the same percentage of TiO$_2$ nanoparticles as in our composite. Comparing (Fig. 4.5) with (Fig. 4.6c), our composite shows a more spherical structure. The absence of HCl may slow down the polymerization, so there are more chances for the interaction of PPy and TiO$_2$ nanoparticles. As discussed above, this spherical structure provides larger surface area than the film. For some applications, such as photocatalytic decomposition of pollutants, this may be the desired morphology as it enhances the contact area with solutions.

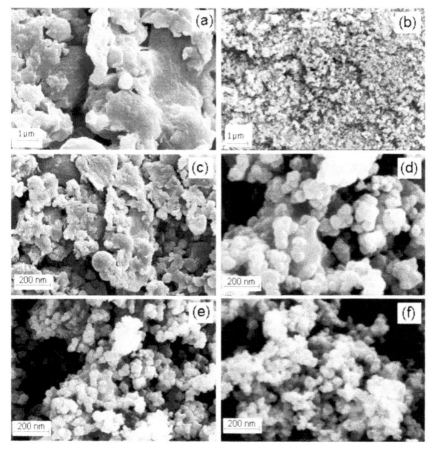

FIGURE 4.6   SEM images of (a) pure PPy, (b) pure TiO$_2$, (c) PPy/TiO$_2$ (0.025), (d) PPy/TiO$_2$ (0.05), (e) PPy/TiO$_2$ (0.075), and (f) PPy/TiO$_2$ (0.1) nanocomposites in the research of Mirzaagha Babazadeh et al. [49] (permission was obtained from the author).

4.4.1.2   TEM

A JEOL 2011 scanning transmission electron microscopy (TEM) was used to obtain TEM micrographs. TiO_2 nanoparticles and polypyrrole/TiO_2 nanocomposites were both dispersed in water. Samples were prepared by evaporation of dispersed sample solution on grids.

TEM micrographs show most TiO_2 nanoparticles are sphere-like structures (Fig. 4.7). Figure 4.8 shows the histogram of TiO_2 nanoparticle size distribution. The mean size is about 4 ± 2 nm. In the hydrothermal method, the average size of TiO_2 nanoparticles can be controlled by adjusting the processing time and temperature [50]. The crystallite size increases with processing time when the nanoparticles diffuse towards each other, aggregate and recrystallize. Average radius of rutile nanoparticles made by hydrothermal method, between 5 and 15 nm, are controlled by diffusion and time (< 6 h), while the size can increase to hundreds of nanometers after 12 h [50]. Temperature can also affect the rate of diffusion, structural coincidences, and recrystallization. By increasing the temperature the diffusion rate increases, so larger nanoparticles can form at higher temperatures [50].

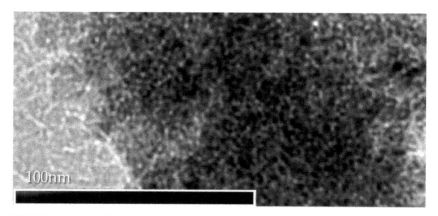

FIGURE 4.7   The TEM micrographs of TiO_2 nanoparticles.

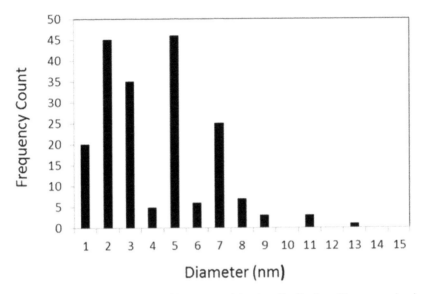

FIGURE 4.8 The histogram of TiO$_2$ nanoparticle size distribution. The mean size is about 4 ± 2 nm.

4.4.2 FTIR SPECTROSCOPY

FTIR spectroscopy is one of the most versatile and efficient techniques to identify the functional groups of PPy [51]. An FTIR spectrum of the sample is displayed in Fig. 4.9. Figure 4.9 shows a strong and broad absorption band at around 3438 cm^{-1}, corresponding to the N–H stretching [52, 53]. These IR spectra should be compared with IR spectra from Luo et al. [52] as our works share the similarities of water as a medium for synthesis and FeCl$_3$ as the oxidant. Our work differs in that we did not use an acid and, more importantly, our TiO$_2$ is in the rutile phase while Luo et al. [49, 52–54] had anatase TiO$_2$. Indeed, all other PPy/TiO$_2$ nanocomposites prepared in water so far have been in the anatase phase; therefore to our knowledge we are reporting the first ever PPy/TiO$_2$ composite in the rutile phase made by chemical oxidation in water.

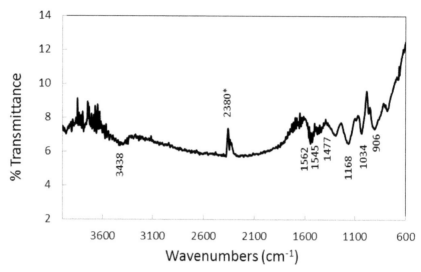

FIGURE 4.9   Infrared spectrum of PPy/TiO2 nanocomposite *CO₂ and N₂ artifact.

Our TiO_2 core is significantly smaller (4 nm compared to 14.6 nm). The N–H stretching frequency in PPy and its composites is in the range of 3300 to 3500 cm⁻¹ [53, 55]. Considering the differences in the phase and the core size, it seems the N–H functional group is not significantly affected by the phase and size of the core, which means the force constant (and therefore the bond strength) for N–H stretching, is not sensitive to those parameters. However, the frequency shift (to significantly larger frequencies) compared to PPy [53, 55] suggests there is a significant binding at the N–H site of the PPy to TiO_2 nanoparticles. This binding is more sensitive to the molecular nature of TiO_2 than to the size or the crystal structure.

Our TGA results (described in Section 4.4.3) also confirm a strong binding of TiO_2 to PPy. The weak peaks from 1656 to 1707 cm⁻¹ might be related to a carbonyl or hydroxyl group in the polymer when overoxidation occurs during polymerization, or possible interaction of PPy and TiO_2 [56]. The weak peaks between 1460 and 1562 cm⁻¹ correspond to the typical pyrrole ring vibration, N–H vibration, C–N vibration and C=C stretching. Among those peaks, the peaks at 1477 and 1545 cm⁻¹ can be assigned to C–N and C=C asymmetric and symmetric ring stretching, respectively [57]. The N–H in-plane mode exhibits a peak at 1168 cm⁻¹ [58]. The peak at 906 cm⁻¹ reflects the C–H out-of-plane deformation vibration [59].

4.4.3 TGA ANALYSIS

The TGA diagrams for pure PPy and the PPy/TiO$_2$ composite are shown in Fig. 4.10. The comparison of pure PPy and PPy/TiO$_2$ composite shows that the mass loss of PPy is more than the composite up to 450°C. The better stability in this range of temperature (from room temperature to 450°C) is due to the presence of TiO$_2$ nanoparticles. It should be noted that our composite has only 2.5% (mass) TiO$_2$.

The degradation of PPy shows a three-stage decomposition pattern as three major slope changes are observed. At $T <$ 100°C, the mass loss is caused by the presence of residual water in the sample. The next stage of the mass loss is by degradation that lasts until 225°C, and is attributed to the loss of dopant ions that are weakly (electrostatically) bound, from the interchain sites of the polymer [37]. When the temperature increases, the more obvious degradation begins. This mass loss is due to degradation and decomposition of the polymer backbone. Unlike pure PPy, the degradation process of PPy/TiO$_2$ composites shows two stages, and the mass loss is less at $T <$ 450°C. TiO$_2$ particles have a positive effect on the degradation in this temperature range which means binding of PPy to TiO$_2$ makes the PPy chemical bonds stronger.

In Curve b, the mass loss of the PPy/TiO$_2$ composite observed at $T <$ 125°C is because of the evaporation of residual water in the sample. The binding of PPy to TiO$_2$ is most probably from the C-NH side (see the discussion of IR spectra in Section 4.4.1) and therefore less NH is available to absorb water (via hydrogen bonding). Another potential cause is the steric effect that could happen due to possible crosslinking of PPy chains on the surface of TiO$_2$ nanoparticles. Therefore there is less mass loss in the PPy/TiO$_2$ composite at $T <$ 125°C. In the second stage from 125°C to 750°C weight loss is because of the degradation and decomposition of the PPy backbone. This decomposition is slower up to 450°C in the PPy/TiO$_2$ composite compared to pure PPy. When the temperature approaches 800°C, the mass loss of PPy/TiO$_2$ is about 90%, while the pure PPy is completely degraded. This is due to the TiO$_2$ content (2.5%) and the residues bound to TiO$_2$ which are stabilized by TiO$_2$ nanoparticles. The conclusion that emerges from comparison of the TGA data of PPy/TiO$_2$ and PPy as well as DSC data that will be discussed later is the following: PPy is more like a linear polymer but when the polymer is grown on the surface of rutile TiO$_2$ nanoparticles a large degree of crosslinking happens. This causes the shift of decomposition of PPy to larger temperatures. At lower temperatures the decomposition is less but at higher temperatures (after the cross linked network is broken) the decomposition is faster, however, this decomposition is limited to a certain mass that includes TiO_2 and small oligomers of pyrrole stabilized by TiO_2, and the content of TiO_2 in the composite (expected to be only 2.5%) (Fig. 4.10).

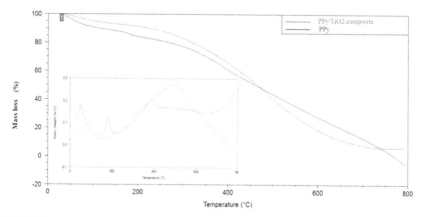

FIGURE 4.10 TGA traces of (a) pure PPy (red curves), (b) PPy/TiO_2 (green curves) composite (3% TiO_2). The derivative plot is shown in the onset.

4.4.5 UV-VIS SPECTRA

We studied the UV-vis absorption spectra of PPy/TiO_2 composites (Fig. 4.11). The spectrum shows three bands at 275 nm, 372 nm, and 620 nm. These three bands can be attributed to pyrrole oligomers and PPy. π-electrons absorb ultraviolet energy to excite these electrons to higher anti-bonding molecular orbitals, so the more easily excited the electrons (i.e., the lower the energy gap between the HOMO and the LUMO), the longer the wavelength. The relation between energy and wavelength can be illustrated by the equation: $E = hc/\lambda$, so the energy of the band gap of PPy is 2.0 eV, which is lower than the band gap energy of rutile TiO_2 (3.02 eV) [60]. Therefore the PPy/TiO_2 composites have potential applications as photoelectric materials, like photovoltaic cells.

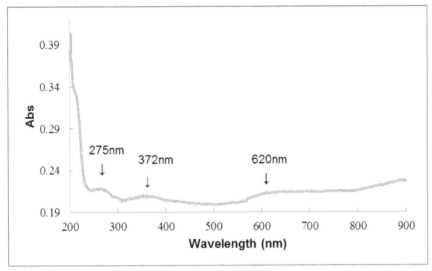

FIGURE 4.11   UV-vis absorption spectrum of PPy/TiO₂ composite. Three bands at 275 nm and 372 nm and 620 nm are assigned to pyrrole oligomers and polypyrrole bound to TiO₂.

Because the PPy composite is slightly dissolved, the UV-vis peaks of PPy are very weak. With different polymer chains, the peaks' positions are also shifted. Figure 4.12 shows the UV-vis absorption spectra of PPy/TiO₂ composites studied by Shengying Li et al. [53] The UV-vis spectrum of PPy/TiO₂ shows two absorption peaks at 335 and 600 nm, respectively. Comparing their absorption peaks to ours, the absorption wavelengths are shorter, which shows that the PPy synthesized in our composites has longer polymer chains. This may mean that the phase and/or the dimension of the core TiO₂ nanoparticles could affect the polymerization process on the surface of TiO₂ nanoparticles. We are doing further investigations to establish the effects of both phase and dimension on the polymerization process.

As discussed above, PPy has smaller band gap than TiO₂ nanoparticles, so the interaction of PPy and TiO₂ nanoparticles in the composites makes the band gap of the new composites adjustable, which makes them widely applicable. Jiao Wang et al. [61] reported that the lowest unoccupied molecular orbital (LUMO) level of the PPy is higher in energy than PPy/TiO₂ electrodes, and hence provides a stronger driving force for the electron injection to the conduction band (CB) of TiO₂, resulting in the improvement of the photoresponse (Fig. 4.13). Upon illumination, the electrons in PPy are excited to the conduction band. Since the LUMO level of TiO_2 is lower in energy than that of PPy, a strong driving force is generated which injects the excited electrons to the conduction band of TiO_2, while the holes are left in the PPy. The electron injection is important to generate the photocurrents [62].

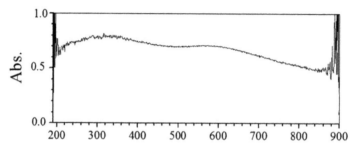

FIGURE 4.12   UV-vis absorption spectrum of PPy/TiO_2 composite from the paper [53] (Reproduced with permission from Journal of Materials Research, 2009, 24, P. 2547–2554. Copyright © Materials Research Society 2009).

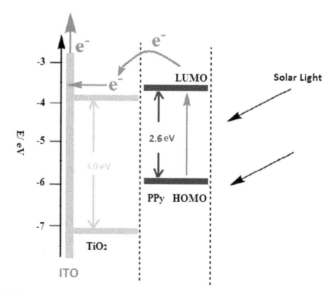

FIGURE 4.13   Schematic diagram of the energy levels of the PPy-TiO_2 electrode [61] representing the possible mechanism of enhancement of TiO_2 photocatalytic properties when embedded in the PPy.

4.4.6   DIFFERENTIAL SCANNING CALORIMETRY (DSC)

Figure 4.14 shows the DSC plot of PPy/TiO$_2$ composite. As mentioned earlier in the TGA curve (Fig. 4.10), there is a small mass loss due to the evaporation of residual water. Now referring to the DSC plot, there is a wide endothermic part in the first heating curve of PPy (Fig. 4.14). This peak could be related to water vaporization. The broad peak could be a result of porous structure. The large porous network may make it harder to remove bound water deep within the composite. The glass transition phase is not apparent in the composite sample (Fig. 4.14).

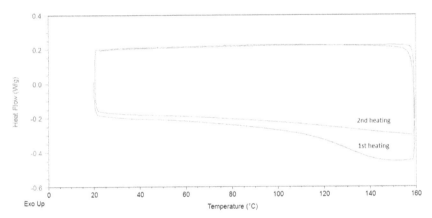

FIGURE 4.14   DSC curves of PPy/TiO$_2$ composite, showing the curves of two cycles.

Figure 4.15 shows the DSC curves of pure PPy at the second and third heating and cooling cycles. The temperature range is from 20°C to 150°C at a rate of 20°C/min. There are no obvious phase transition peaks, but there is a small step at 105°C during the cooling process (Fig. 4.16), which may be ascribed to the glass transition. Since only 2.15 mg PPy was used for DSC, the glass transition step is small.

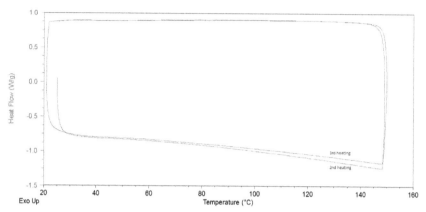

FIGURE 4.15   DSC plot of prepared PPy, showing the second and third heating cycles.

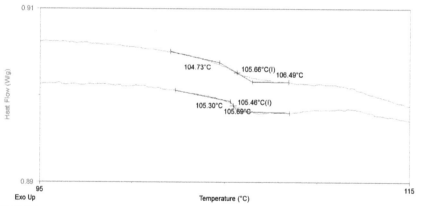

FIGURE 4.16   The amplification of cooling curves of pure PPy at the temperature between 95°C and 115°C (the bottom DSC trace). There is a small step at around 105°C (the top DSC traces). This may be ascribed to the glass transition temperature.

Some works [58, 63] suggest the glass transition temperature is around 100°C because of the appearance of endothermic dip during heating, but in those studies the heating and cooling cycles were not repeated. Thus their results are not conclusive because the endothermic dip can be caused by the evaporation of moisture in PPy. In the case of our data, there is a wide dip at the temperature between 115°C and 155°C in the first cycle of heating process (Fig. 4.14), but not observable in the second cycle. Among the papers, which reported the DSC curves with repeated heating cycles, the glass transition temperature is still controversial due to the complicated structure of PPy caused by the defects such as crosslinking and overoxidation. Some researches stated that there is no glass transition [64], while some other researchers reported the glass transition temperature of PPy in a range of 71°C to 155°C [65–67].

Figure 4.17 also shows the DSC traces of PPy/TiO$_2$ with a similar relative y-scale to the PPy. Clearly there is no glass transition despite using four times more sample. This could be due to the extensive cross-linking on the surface of TiO$_2$. It can also be partially responsible for less water absorption and other differences discussed in the examination of TGA results for the nanocomposite and PPy.

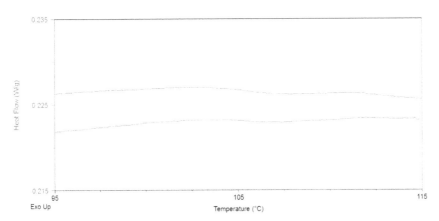

FIGURE 4.17   The amplification of cooling curves of pure PPy/TiO$_2$ composite at the temperature between 95°C and 115°C (the bottom DSC trace).

4.5   CONCLUSIONS

In this research, we synthesized PPy/TiO$_2$ composites in aqueous solution by one-step polymerization. We successfully prepared core-shell-structured PPy/TiO$_2$ nanoparticles with PPy on the surface of TiO$_2$ nanoparticles. The PPy/TiO$_2$ nanocomposites have higher thermal stability, up to 450°C, than the pure PPy. In order to decrease the size of nanocomposites, an ultrasonic bath was applied during the suspension of TiO$_2$ nanoparticles, inhibiting the agglomeration of nanoparticles. The glass transition in the PPy/TiO$_2$ composite does not exist. We attributed this along with the

TGA results to the increased crosslinking of PPy on the surface of TiO₂ Nanoparticles.

The nanocomposites have potential applications as photovoltaic materials and photocatalysts. The UV-V is peaks shift to longer wavelengths; this is an advantage of the nanocomposites made with the rutile phase. Despite the change in reactivity of pyrrole oligomers on the surface of the rutile and Anatase phases, the bond energies at the binding site (likely to be NH) are not significantly dependent on phase or nanoparticle size. The PPy on the surfaces of rutile phase grows to longer chains when compared to Anatase phase. This is depicted in the Fig. 4.18.

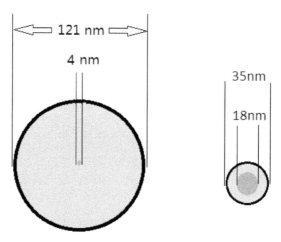

FIGURE 4.18   The core-shell structures when rutile (large shell, small core) and Anatase (small shell, large core) TiO₂ are used to make nanocomposites.

ACKNOWLEDGMENTS

This research was financially supported by the Natural Sciences and Engineering Research Council of Canada, and the New Brunswick Innovation Foundation. K.G. acknowledges the Canada Foundation for Innovation, Atlantic Innovation Fund and other partners that fund the Facilities for Materials Characterization managed by the Institute for Research in Materials at Dalhousie University. We thank James M. Ehrman, Dr. Ralf Brüning and Tanu Sharma from Mount Allison University for their kind help with the SEM images and XRD data.

KEYWORDS

- **Metal nanoparticles**
- **Nanocomposite**
- **Oxidative polymerization**
- **Photocatalysis**
- **Polypyrrole**
- **TiO$_2$**

REFERENCES

1. Hakim, S. H. & Shanks, B. H. (2011). Synthesis and characterization of hierarchically structured alumino silicates. Journal of Materials Chemistry, 21(20), 7364–7375.
2. Shirakawa, H. et al. (1977). Synthesis of electrically conducting organic polymers, halogen derivatives of polyacetylene, (CH)x. Journal of the Chemical Society, Chemical Communications, 16, 578–580.
3. Takahashi, R. et al. (2005). Effect of diffusion in catalytic dehydration of alcohol over silica–alumina with continuous macropores. Journal of Catalysis, 229(1), 24–29.
4. Burns, F., Themens, P., & Ghandi, K. (2012). Assessment of solubility of cellulose in phosphonium ionic liquids and their mixtures with DMF. Composite Interfaces, (submitted).
5. Ghandi, K., & Greenway, K. (2010). Process for the production of polystyrene in phosphonium ionic liquids, in WO Patent WO/2010/108, 271.
6. Ghandi, K., & Themens, P. (2011). Novel magnetic nanocomposite material and processes for the production thereof, in US Patent, 61–491557.
7. Ghandi, K. (2012). Process for the production of polystyrene and novel polymers in phosphonium ionic liquids, in international patent. US Patent, 20–120–049–101.
8. Ghandi, K. et al. (2012). A Novel Polymer solar cell, in US Patent 2012–61–662546.
9. Mitchell, G. R., Davis, F. J., & Legge, C. H. (1988). The effect of dopant molecules on the molecular order of electrically conducting films of polypyrrole, Synthetic Metals, 26(3), 247–257.
10. Bredas, J. L. et al. (1984). The role of mobile organic radicals and ions (solitons, polarons and bipolarons) in the transport properties of doped conjugated polymers. Synthetic Metals, 9(2), 265–274.
11. Brédas, J. L. et al. (1984). Polarons and bipolarons in polypyrrole, Evolution of the band structure and optical spectrum upon doing. Physical Review B, 30(2), 1023–1025.
12. Skotheim, T. A., & Reynolds, J. R. (2006). Hand book of conducting polymers, Recent advances in polypyrrole, ed. Cho, S. H., Song, K. T., & Lee, J. Y. CRC Press.

13. DeArmitt, C. (1995). Novel colloidal and soluble forms of polyaniline and polypyrrole, University of Sussex.

14. Diaz, A. F., Kanazawa, K. K., & Gardini, G. P. (1979). Electrochemical polymerization of pyrrole. Journal of the Chemical Society, Chemical Communications, 14, 635–636.

15. Kanazawa, K. K. et al. (1979). 'Organic metals,' polypyrrole, a stable synthetic 'metallic' polymer Journal of the Chemical Society, Chemical Communications, 19, 854–855.

16. Diaz, A. F. (1981). Electrochemical preparation and characterization of conducting polymers. Chemica Scripta, 17, 145–148.

17. Ateh, D. D., Navsaria, H. A., & Vadgama, P. (2006). Polypyrrole-based conducting polymers and interactions with biological tissues. Journal of the Royal Society Interface, 11(3), 741–752.

18. Wei, Y. et al. (2010). One-step UV-induced synthesis of polypyrrole/Ag nanocomposites at the water/ionic liquid interface. Nanoscale Research Letters, 5, 433–437.

19. Wang, J., Neoh, K. G., & Kang, E. T. (2004). Comparative study of chemically synthesized and plasma polymerized pyrrole and thiophene thin films. Thin Solid Films, 446(2), 205–217.

20. Karim, M. R., Lee, C. J., & Lee, M. S. (2007). Synthesis of conducting polypyrrole by radiolysis polymerization method. Polymers for Advanced Technologies, 18, 916–920.

21. Kumar, D., & Sharma R. C. (1998). Advances in conductive polymers. European Polymer Journal,. 34(8), 1053–1060.

22. Ruckenstein, E., & Chen, J. H. (1991). Polypyrrole conductive composites prepared by coprecipitation. Polymer, 32(7), 1230–1235.

23. Kudoh, Y. (1996). Properties of polypyrrole prepared by chemical polymerization using aqueous solution containing $Fe(SO_4)_3$ and anionic surfactant. Synthetic Metals, 79, 17–22.

24. Wang, D. et al. (2010). Spontaneous Growth of Free-Standing Polypyrrole Films at an Air/Ionic Liquid Interface. Langmuir, 26(18), 14405–14408.

25. Myers, R. (1986). Chemical oxidative polymerization as a synthetic route to electrically conducting polypyrroles, Journal of Electronic Materials, 15(2), 61–69.

26. Vork, F. T. A., & Janssen, L. J. J. (1988). Structural effects in polypyrrole synthesis. Electrochimica Acta. 33(11), 1513–1517.

27. Calvo, P. A. et al. (2002). Chemical oxidative polymerization of pyrrole in the presence of m-hydroxybenzoic acid and m-hydroxycinnamic acid-related compounds, Synthetic Metals, 126(1), 111–116.

28. Cao, F., Oskam, G., & Searson, P. C. (1995). A Solid State, Dye Sensitized Photoelectrochemical Cell, The Journal of Physical Chemistry, 99(47), 17071–17073.

29. Dutta K., & De, S. K. (2006). Transport and optical properties of SiO2-polypyrrole nanocomposites, Solid State Communications, 140(3–4), 167–171.

30. Holister, P. et al. (2003). Nanoparticles: technology white papers, 3, 2–11.

31. Chen, X., & Mao, S. S. (2007). Titanium Dioxide Nanomaterials, Synthesis Properties, Modifications and Applications, Chemical Reviews, 107, 2891–2959.

32. Xu, S. H. et al. (2010). Improving the photocatalytic performance of conducting polymer polythiophene sensitized TiO$_2$ nanoparticles under sunlight irradiation, Reaction Kinetics, Mechanisms and Catalysis, 101, 237–249.

33. Mital, G. S., & Manoj, T. (2011). A review of TiO2 nanoparticles, Chinese Science Bulletin, 56, 1639–1657.

34. Baldassari, S., & Komarneni, S. (2005). Rapid microwave-hydrothermal synthesis of anatase form of titanium dioxide, Communications of the American Ceramic Society, 88(11), 3238–3240.

35. Landmann, M., & Rauls, E. W. G. S. (2012). The electronic structure and optical response of rutile, anatase and brookite TiO$_2$, Journal of Physics, Condensed Matter, 24, 195503.

36. Gangopadhyay, R., & De, A. (2000). Conducting Polymer Nanocomposite, A Brief Overview, Chem. Mater, 12, 608.

37. Kumar, A., & Sarmah, S. (2011). AC conductivity and dielectric spectroscopic studies of polypyrrole-titanium dioxide hybrid nanocomposites, physica status solidi (a). 208(9), 2203–2210.

38. Armes, S. P. et al. (1991). Conducting polymer-colloidal silica composites, Polymer, 32(13), 2325–2330.

39. Bhattacharya, A. et al. (1996). A new conducting nanocomposite-PPy-zirconium (IV) oxide, Materials Research Bulletin, 31(5), 527–530.

40. Cheng, Q., et al. (2006). Synthesis and structural properties of polypyrrole/nano-Y$_2$O$_3$ conducting composite, Applied Surface Science, 253(4), 1736–1740.

41. Trung, V. Q., Tung, D. N., & Huyen, D. N. (2009). Polypyrrole/Al$_2$O$_3$ nanocomposites: preparation, characterization and electromagnetic shielding properties, Journal of Experimental Nanoscience, 4(3), 213–219.

42. Senadeera, G. K. R. et al. (2006). Enhanced photoresponses of polypyrrole on surface modified TiO$_2$ with self-assembled monolayers, Journal of Photochemistry and Photobiology A Chemistry, 184, 234.

43. Cervini, R., Cheng, Y., & Simon, G. (2004). Solid-state Ru-dye solar cells using polypyrrole as a hole conductor, Journal of Physics D, Applied Physics, 37, 13.

44. Vermeir, I. E., Kim, N. Y., & Laibinis, P. E. (1999). Electrical properties of covalently linked silicon/polypyrrole junctions, Applied Physics Letters, 74, 3860.

45. Kubisa, P. (2009). Ionic liquids as solvents for polymerization processes-Progress and challenges, Progress in Polymer Science, 34, 1333–1347.

46. Cassaignon, S., Koelsch, M., & Jolivet, J. P. (2007). Selective synthesis of brookite, anatase and rutile nanoparticles: thermolysis of TiCl4 in aqueous nitric acid, Journal of Materials Science, 42, 6689–6695.

47. Luo, M. et al. (2010). Synthesis and Structural and Electrical Characteristics of Polypyrrole Nanotube/TiO2 Hybrid Composites, Journal of Macromolecular Science, Part B, 49(3), 419–428.

48. Liu, J., & Wan, M. (2001). Synthesis Characterization and Electrical Properties of Microtubules of Polypyrrole Synthesized by a Template-Free Method. Journal of Materials Chemistry, 11(2), 404–407.

49. Babazadeh, M., Gohari, F. R., & Olad, A. (2012). Characterization and Physical Properties Investigation of Conducting Polypyrrole/Tio2 Nanocomposites Prepared through a one-step "in situ" Polymerization Method, Journal of Applied Polymer Science, 123(4), 1922–1927.

50. Reyes-Coronado, D. et al. (2008). Phase-pure TiO_2 Nanoparticles, Anatase, Brookite and Rutile, Nanotechnology, 19(14), 145605.

51. Toshima, N., & Ihata, O. (1996). Catalytic Synthesis of Conductive Polypyrrole using Iron(III) Catalyst and Molecular Oxygen, Synthetic Metals, 79, 165–172.

52. Luo, Q. et al. (2011). Photocatalytic Activity of Polypyrrole/TiO2 Nanocomposites under Visible and UV light, Journal of Materials Science, 46(6), 1646–1654.

53. Li, S. et al. (2009). Preparation and characterization of polypyrrole/TiO2 nanocomposite and its photocatalytic activity under visible light irradiation, Journal of Materials Research, 24(8), 2547–2554.

54. Wang, D. et al. (2008). Sunlight Photocatalytic Activity of Polypyrrole-TiO2 Nanocomposites Prepared by 'in situ' method, Catalysis Communications, 9(6), 1162–1166.

55. Davidson, R. G., & Turner, T. G. (1995). An IR Spectroscopic Study of the Electrochemical Reduction of Polypyrrole Doped with Dodecylsulfate Anion, Synthetic Metals, 72(2), 121–128.

56. Hahn, S. J. et al. (1986). Auger and Infrared Study of Polypyrrole Films, Evidence of Chemical Changes During Electrochemical Deposition and Aging in Air, Synthetic Metals, 14(1–2), 89–96.

57. Kato, H. et al. (1991). Fourier Transform Infrared Spectroscopy Study of Conducting Polymer Polypyrrole, Higher Order Structure of Electrochemically Synthesized Film. The Journal of Physical Chemistry, 95(15), 6014–6016.

58. Vishnuvardhan, T. K. et al. (2006). Synthesis, characterization and AC conductivity of Polypyrrole/Y_2O_3 Composites. Bulletin of Materials Science, 29(1), 77–83.

59. Chougule, M. A. et al. (2011). Synthesis and Characterization of Polypyrrole (PPy) Thin Films, Soft Nanoscience Letters, 1, 6–10.

60. Gupta, S., & Tripathi, M. (2011). A Review of TiO_2 Nanoparticles, Chinese Science Bulletin, 56(16), 1639–1657.

61. Wang, J., & Ni, X. (2008). Photoresponsive Polypyrrole-TiO_2 Nanoparticles Film Fabricated by a Novel Surface Initiated Polymerization. Solid State Communications, 146 239–244.

62. Cahen, D. et al. (2000) Nature of Photovoltaic Action in Dye-Sensitized Solar Cells, The Journal of Physical Chemistry B, 104(9), 2053–2059.

63. Mavinakuli, P. et al. (2010). Polypyrrole/Silicon Carbide Nanocomposites with Tunable Electrical Conductivity, The Journal of Physical Chemistry C, 114(9). 3874–3882.

64. Ozkazanc, E. et al. (2012). Preparation and Characterization of Polypyrrole/selenium Composites, Polymer Engineering and Science, 23363.

65. SANDU, T. et al. (2012). Characterization of Functionalized Polypyrrole, Revue Roumaine de Chimie, 57(3), 177–185.

66. Truong, V. T., Ennis, B. C., & Forsyth, M. (1995). Enhanced Thermal Properties and Morphology of Ion-exchanged Polypyrrole films, Polymer, 36(10), 1933–1940.

67. Abbasi, A. M. R., Marsalkova, M., &. Militky, J. (2013). Conductometry and Size Characterization of Polypyrrole Nanoparticles Produced by Ball Milling, Journal of Nanoparticles, 2013, 4.

CHAPTER 5

POLY (LACTIC ACID) BASED HYBRID COMPOSITE FILMS CONTAINING ULTRASOUND TREATED CELLULOSE AND POLY (ETHYLENE GLYCOL) AS PLASTICIZER AND REACTION MEDIA

KATALIN HALÁSZ, MANDAR P. BADVE, and LEVENTE CSÓKA

CONTENTS

Abstract 102
5.1   Introduction 102
5.2   Experimental 104
5.3   Results and discussions 106
5.4   Conclusions 116
Keywords 117
References 117

ABSTRACT

Poly (lactic acid) (PLA) based composite films were prepared by melt blending of PLA and microcrystalline cellulose (MCC) ultrasound treated or dispersed in poly (ethylene glycol) (PEG400). To evaluate the effectiveness of the ultrasound (US) treatment and to characterize the extruded PLA based films tensile testing, differential scanning calorimetry (DSC), thermo gravimetric analysis (TGA), scanning electron microscopy (SEM), wide angle X-ray diffraction (WAXD) were carried out. From the results it can be concluded that the PEG400 can be a good media for reducing the cellulose particle size by ultrasonic treatment. The cellulose nanocrystals produced by US treatment were spherical like in shape and although there still remained particles above the nano-range, thus hybrid composites were prepared, significant improvement occurred in the toughness of PLA.

5.1 INTRODUCTION

Poly (lactic acid) or PLA is biobased thermoplastic polyester, which can be produced from lactic acid derived from the fermentation of different naturally available polysaccharides. Furthermore, PLA is a biodegradable and compostable plastic with relative good properties compared to the other biodegradable, biobased plastics, thus PLA has got its potential in many applications such as in medical, drug delivery, textile or packaging applications. However, PLA is too brittle for many applications, softens at relative low temperatures, and has weak water vapor and gas barrier properties compared to commercial polymers. To extend the application field the improvement of its properties (barrier, thermal, mechanical) is required.

There have been taken many efforts to improve the week properties of PLA with different kind of materials including blending/compounding with other polymers, plasticizers [1–10], reinforcing materials in micro (e.g., natural fibers or particles [11–15]) and nanosized (layered silicates, carbon nanotubes, nanoparticles or nanocrystals) [16–26]. One of the most promising materials to develop the properties of the PLA is cellulose. Cellulose is the most abundant renewable, biodegradable polymer on the Earth and its properties like low cost, low density, high specific strength, high modulus and relatively high surface [27, 28] give a widespread industrial usage. Nevertheless, using cellulose has got also its limitations too since cellulose has a strong sensitivity to water and moisture, when it is dried it may form aggregates and it display poor compatibility with the hydrophobic polymeric matrices [28]. However, if a well-dispersed, homogenous system with good interfacial interactions is created, cellulose can show its benefits.

In the literature there have been already used microcrystalline cellulose [29–31] and microfibrillated cellulose (MFC) [32, 33] to improve the properties of the PLA. Nanosized cellulose like cellulose nanofibers (CNF) [34, 35] and cellulose nanowhiskers (CNW) [35–42] could offer further improvements. Cellulose nanocrystals are mostly prepared in water media with acidic hydrolysis. To remove the media the process is usually followed by lyophilization. However, freeze drying typically results in agglomeration/aggregation of the CNC thus a fine dispersed system is difficult to achieve especially in case of melt processed composites. In order to enhance the dispersibility of the CNC in PLA matrix solvent based composite producing process could be applied or a solvent based masterbatch preparation before the melt process. However, using large amount of solvent is incompatible with the environmentally friendly nature of poly (lactic acid). Surface modification of the CNC with anionic surfactants can be a solution to improve the interfacial interactions between cellulose and PLA as [38, 40, 42] reported. Oksman et al. [37] found that the presence of poly (ethylene glycol) (PEG1500) increased the dispersibility of CNC in PLA matrix too (which was an effective plasticizer for the PLA as well). Qu et al. [34] reported that in case of chemo-mechanical produced CNF PEG1000 seemed to be a good choice in increasing the interactions between the PLA and the cellulose.

According to [30] poly(ethylene glycol) (PEG) can improve the interfacial interactions between PLA and cellulose since the C-O-C and the O-H of the PEG can form H-bonding or dipolar interactions between matrix and reinforcing material. PEG cannot only act as compatibilizer but as plasticizer as well thus reducing the brittleness of the PLA [2, 43, 44]. In general, the lower the molecular weight the higher the plasticizing effect [2, 43, 45] due to PEGs with lower molecular weights have larger number of hydroxyl groups (which can develop hydrogen bonds between the polymer and the plasticizer replacing the polymer-polymer interactions.) per mole compared to the PEGs with higher molecular weights. Low molecular weight PEGs has further advantages namely they are nontoxic, often used as medicine or food additives and they are biodegradable like PLA via microorganisms.

The aim of our study was to investigate the effect of ultrasonication applied on cellulose in PEG400 media on the properties of PLA matrix. Ultrasonication, with the aim of producing cellulose nanocrystals, was carried out in PEG400 media without adding any other chemicals thus keeping the process "green." Since PEG400 is a good plasticizer for PLA, and it can generate good interfacial interactions between cellulose and PLA, it is not required to remove from the system thus using liquid PEG400 allows the excluding of the freeze-drying step as well.

5.2   EXPERIMENTAL

5.2.1   MATERIALS AND METHODS

Poly(lactic acid) transparent, extrusion grade granulate was supplied from Shenzhen Bright China Industrial Co. with trade name Esun™ AI1031, microcrystalline cellulose was obtained from Sigma Aldrich in particle size <20μm. Poly (ethylene glycol) (PEG400) was received from Sigma Aldrich and used in liquid form with the average molecular weight of 400.

To indicate particle size reduction and to help the dispersion of the cellulose MCC-PEG suspension was prepared (in 10, 30 and 50 wt.% to keep the PEG400 content constant in the PLA based composites) and treated with ultrasound. Since the power of direct ultrasonication with the ultrasonic horn (TESLA, 20 kHz) seemed to be too high and it caused quick degradation of the PEG400 the time of the direct ultrasonication was reduced to 15 min and the treatment was continued with indirect sonication in ultrasonic bath (TESLA, with dual-frequency unit - transducers frequencies of 25 and 75 kHz) for 40 min. To obtain nano sized cellulose chemicals were not used in order to keep the process "green." Prior the treatment the microcrystalline cellulose was swelled in PEG400 for 24 h. The swelling process was carried out in case of MCC without ultrasonic treatment as well.

Since hydrolytic degradation of PLA can appear during the extrusion [46], PLA and MCC were properly dried (at 60 °C for 4 day) prior all the used processing technologies. Suspension of microcrystalline cellulose with or without ultrasonic treatment in poly(ethylene glycol) and PLA granulates were mixed in a COLLIN ZK25T four zoned compact laboratory twin screw corotating extruder with a screw speed of 50 rpm The temperature profile varied from 170 °C at the feeding zone to 190 °C at the die. The composite films (Table 5.1) were prepared on a twin-screw extruder (LABTECH Scientific twin screw extruder with melt pump and LBRC-150 chill roll cast line) in the thickness of 90–110 microns. Control samples were prepared through the above-described processes.

TABLE 5.1   List of Characterized Materials, their Compositions and Numbers Shown in Figures

1.	Neat PLA
2.	PLA+10wt%PEG
3.	PLA+1wt% MMC + 10wt% PEG
4.	PLA+3wt% MMC + 10wt% PEG
5.	PLA+5wt% MMC + 10wt% PEG
6.	PLA+1wt% USMMC + 10wt% PEG
7.	PLA+3wt% USMMC + 10wt% PEG
8.	PLA+5wt% USMMC + 10wt% PEG

US: ultrasonically treated.

5.2.2 CHARACTERIZATION

Wide-angle X-ray (WAXD) scattering measurements were carried out with a Philips PW1710 diffractometer. WAXD patterns were obtained at room temperature using nickel-filtered Cu Kα radiation with generator tension of 50 kV, generator current of 40 mA, a wavelength of 0.1544 nm, in range of 5–40°2θ.

Transmission electron microscopy (TEM) of the ultrathin PLA based samples was carried out with Jeol JEM-2000EX transmission electron microscope. The acceleration voltage was 120 keV.

Thermal properties were determined by differential scanning calorimetry (DSC) and thermogravimetric analysis (TGA). DSC was performed with Perkin Elmer DSC 7 calorimeter. Each sample was heated from ambient temperature to 200 °C at a rate of 5 °C/min under inert condition; curves obtained from second heating were analyzed. The degree of crystallinity was calculated from the following equation (Eq. (1)):

$$\chi_c(\%) = 100 \, x \, |\Delta H_m + \Delta H_{cc}|/\Delta H^\circ_m \tag{1}$$

where ΔH_m and ΔH_{cc} is the enthalpy of fusion and crystallization at melting and crystallization temperature, respectively. ΔH°_m is the heat of fusion of a perfect orthorhombic PLA crystal with applying the value of 93.6 J/g. The thermogravimetric analysis (TGA) was performed with Perkin Elmer TG 7. Each sample was heated from ambient temperature to 505 °C at the rate of 5 °C/min under inert condition (in N_2).

Tensile properties of the samples were measured both in cross and pro-duction direction according to EN 527 using Instron 3345 tensile tester. The measurements were performed at 23 °C and 50RH%, with 50 mm gauge length, 2 kN max. Load and 50 mm/min crosshead speed. Young-modulus, stress and strain at peak and at break were determined. All re-sults presented are the average values of five measurements.

In order to investigate the microstructure of the samples scanning elec-tron microscopy (SEM) (Hitachi S-3400N) was used. The secondary elec-tron images were taken of the fracture surfaces. The acceleration voltage was 20 kV and the specimens were coated with gold to avoid charging.

5.3   RESULTS AND DISCUSSIONS

The WAXD diffractograms are shown on the (Fig. 5.1). In the Fig. 5.1A the WAXD pattern of the neat PLA (a) and the MCC (b), on the (Fig. 5.1B) the pattern of the composite materials comparing to the neat PLA are pre-sented. The neat PLA shows four main peaks, which can be corresponded to the different crystal structures. The peak at $2\Theta=16.7°$ (5.304 Å) is the peak of homo-crystal form, while the other peaks ($2\Theta=9.5°$, $16.7°$, $19.5°$, $24.9°$–9.230, 4.564, 3.521 Å, respectively) can be attributed to the ste-reocomplex crystal form (47,48). The broad hump shows amorphous na-ture. In case of MCC the peaks occurred at $2\Theta=9.4°$, $22.4°$, $35.00°$ (5.676, 3.961, 2.598 Å) are corresponding to the cellulose I polymorph structure [29], a board peak appeared as well at $2\Theta=16.5°$. The composites (letters mean the same as in the Table 5.1) show no peaks but broad hump with shoulder (except in case of PLA containing 3 wt.% MCC and 10 wt.% of PEG where a little peak can be observed at $2\Theta=16.6°$). Mathew et al. [29] explains this phenomenon to occur due to the fast cooling rates during extrusion, which indicate low crystallinity. The diffractograms show regu-lar trend in the intensity of the shoulder (later board peak), the higher the

MCC content the higher the peak formed around at $2\Theta=22.4°$, moreover higher intensity appeared when ultrasound treatment was used, probably due to the higher crystallinity of cellulose caused by the treatment, which degraded some of the amorphous regions in the MCC [49]. The intensity of the peak (which can be originated mainly from the PLA crystallinity but from the MCC as well) around $2\Theta=16.7°$ on the other hand do not show any regular trends in intensity change.

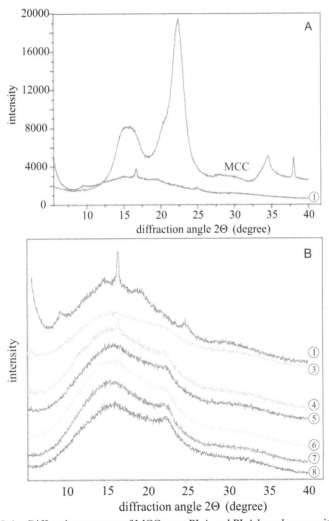

FIGURE 5.1   Diffraction patterns of MCC, neat PLA and PLA based composites.

After cutting ultrathin films of the composites transmission electron microscopy (TEM) was carried out. In Fig. 5.2, TEM images of composites containing MCC and PEG without ultrasonication can be seen. Because of the large particles it was difficult to get a true view, but as the (Fig. 5.2) shows the extrusion process itself may cause cellulose particles size reduction. Ultrasonically treated cellulose particles can be seen in the poly (lactic acid) matrix in Fig. 5.2 as well. The less the concentration of the MCC-PEG suspension was the more effective the US treatment became. There are small cellulose crystals under 100 nm in the PLA based composite containing USMCC. The cellulose crystals created through ultrasonication in PEG400, just as [49] observed too after ultrasonic treatment of cellulose in water media, have spherical shape instead of whisker. Because of the sensitivity of the PLA to the relative high acceleration voltage closer images could not be taken.

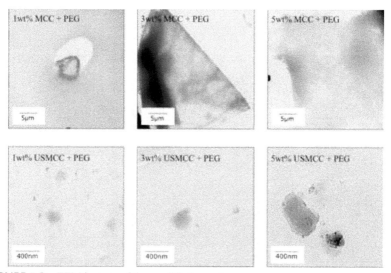

FIGURE 5.2   TEM images of the modified samples.

DSC measurement (Fig. 5.3 and Table 5.2) was used to characterize the thermal properties of neat PLA and PLA based composites. Glass transition temperatures (T_g) of the modified materials were decreased or cannot be observed due to the plasticizer, which provided more flexibility to the polymer chains. In the DSC heating scans all PLA and PLA based samples showed exothermic peak, which attributed to the cold crystallization (T_{cc}).

The amorphous part of the modified poly (lactic acid) samples started to form organized structure at lower temperatures (T_{cc}). Due to the presence of PEG enhanced chain mobility occurred in the amorphous phase and PLA crystallizes with more ease at lower temperatures. The exothermal cold crystallization peaks are smaller and thinner compared to the neat PLA, but as one can see in some cases the melting peaks are wider indicating higher crystallinity. The melting temperatures did not show notable difference except the difference of the two melting peaks. While the neat PLA bears two well-defined individual peaks, the first melting peaks of the composites are not so strong and they shifted to lower temperatures. The two melting peaks of neat PLA can be due to the coexistence of two kinds of crystalline structure, or because of the melting behavior with melt recrystallization model [50, 51] or due to the dual lamellae population [51]. The smaller peaks can indicate that the plasticizer and the additives changed the crystal structure of the PLA (in case of Ref. [7] no other peak can be observed). According to the enthalpies and the calculated crystallinity the cellulose may act as nucleating agent especially when ultrasonic treatment was used. According to Ref. [52], it is reported that the nucleating effect is enhanced if homogenous cellulose dispersion in poly (lactic acid) matrix is achieved. This indicates that more homogenous structure was obtained with US treatment, which modified the cellulose characteristics as well and caused significant size reduction.

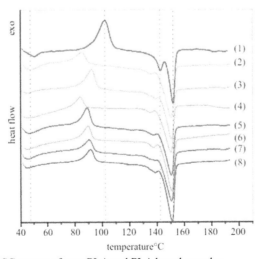

FIGURE 5.3 DSC curves of neat PLA and PLA based samples.

TABLE 5.2   Thermal Properties Measured by DSC

	Tm melt peaks (°C)		Tg (°C)	Tcc(°C)	ΔHm(J/g)	ΔHcc(J/g)	χc (%)
1.	143.5	151.6	43.9	102.8	33.76	21.41	13.19
2.	134.2	150.1	41.2	97.3	31.46	7.95	25.12
3.	138.6	150.7	40.9	92.8	28.29	11.13	18.33
4.	135.4	149.9	41.5	83.4	27.48	6.64	22.26
5.	137.2	150.1	42.0	88.6	28.74	11.30	18.63
6.	137.8	150.3	41.1	89.7	28.89	9.17	21.07
7.	–	150.5	42.3	90.3	29.24	7.52	23.21
8.	138.6	150.5	42.5	91.6	30.11	5.59	26.20

The thermogravimetric (TGA) and derivative thermogravimetric (DTG) curves of neat PLA and PLA based composites are shown in Figs. 5.4 and 5.5, the values for onset temperature of the thermal degradation (T_{ons}) and temperature of maximal rate of weight loss (T_{max}) are listed in Table 5.3. In almost every cases the onset temperatures of the composites shifted to higher temperatures even if only PEG400 was used as modifier. Rodríguez-Llamazares et al. [23] reported similar change in T_{ons} adding PEG3350 to PLA matrix. The highest improvement (with 8, 8 °C) appeared when US treated cellulose in 5 wt.% was incorporated into the PLA. According to the DTG curves the samples degraded in one step (neat PLA showed one step also but with a complex decomposition) which is probably due to that the cellulose thermal degradation appears at the same temperatures as the PLA, but the one single peak can indicate that there is strong hydrogen bonding between cellulose and poly (lactic acid) matrix as well [45]. All the T_{max} peaks shifted to lower values, the samples started to decompose at lower temperatures.

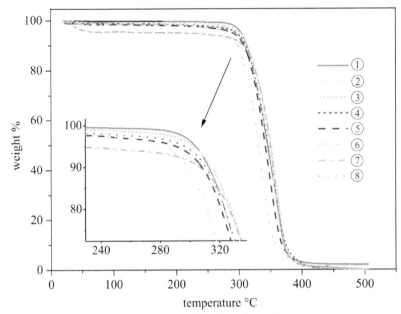

FIGURE 5.4   TG curves of neat PLA and PLA based samples.

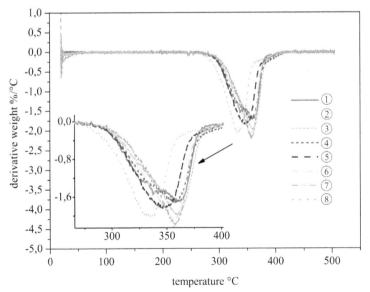

FIGURE 5.5   DTG curves of neat PLA and PLA based samples.

TABLE 5.3   Values of Tmax and Tonset

	Tmax	Tonset
1.	315.3	360.0
2.	318.0	357.9
3.	316.1	360.0
4.	313.1	357.6
5.	316.3	347.2
6.	319.0	356.9
7.	306.3	334.1
8.	324.1	358.2

Stress-strain curves of the neat and the modified samples in machine and in cross directions are shown in Figs. 5.6 and 3.7. According to the results neat PLA showed no necking, it had broken with very little elongation, it exhibited no or little plastic deformations. On the other hand most of the modified samples containing PEG400 and cellulose showed plastic deformation with yielding. The maximal improvement at machine direction in strain at peak was 106% (strain was 4.2%) in case of PLA containing 1 wt.% USMCC with PEG, at cross section sample with the same composition showed the highest improvement (113%), where the strain was 4.178%. More remarkable improvement was observed in case of strain at break, where samples with US treated MCC showed significant enhancement at machine and cross directions as one can see in Figs. 5.6 and 5.7. The strains at break in the machine direction are 303.5, 236.0% and 183.4% in case of using 1, 3 and 5 wt.% of USMCC with 10 wt.% of PEG, respectively. The neck which was formed during strain softening extended along these samples and large amount of cold drawing occurred with little strain hardening. The high strain-to-failure and the high toughness was mainly due to void formation which was indicated by stress-whitening [53, 54] (Fig. 5.8). The microvoids release the plastic constraint in the matrix, triggering large-scale plastic deformations [54]. The incorporated particles cavitate in the matrix that adsorbs higher energies [55], which resulted in improvement of tensile energy absorption (Fig. 5.9). The tensile energy absorption was enhanced almost in every case (excluded samples containing 1 and 5 wt.% MCC and PEG400 without ultrasound treatment). The explanation of the higher strain and energy absorption of the samples modified with ultrasound treated cellulose lies on the fact that the treatment resulted in fine particles formation. The smaller cellulose particles have got higher surface area, which provides the possibility of better adhesion in the interface. Although strain at break was improved, stress at peak, ultimate strength and Young-modulus were reduced as expected.

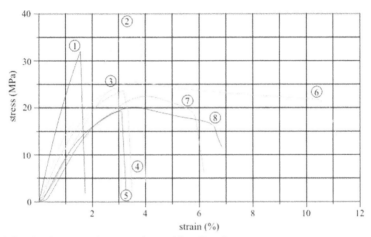

FIGURE 5.6   Stress-strain curves in machine direction.

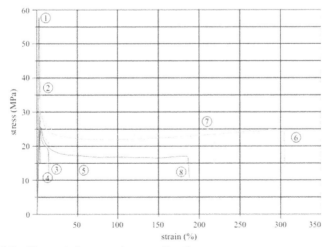

FIGURE 5.7   Stress-strain curves in cross direction.

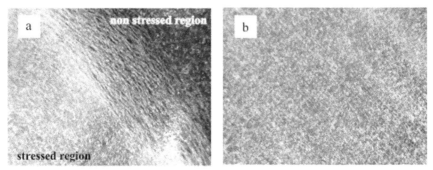

FIGURE 5.8   Optical microscopic images of the stress-whitening, (a) comparison of nonstressed and stressed regions, (b) closer view of the micro void formation.

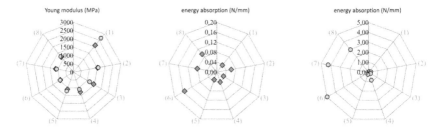

FIGURE 5.9   Young modulus and tensile energy absorption in machine (○) and in cross (□) directions.

Scanning electron microscopic (SEM) images (Figs. 5.10–5.12) were taken of the facture surfaces after the tensile tests. As the images demonstrate the neat PLA shows rigid broken surface, no plastic regions can be observed (Fig. 5.10). In contrast sample containing PEG 400 shows plastic deformations as well as the other modified samples (Figs. 5.10–5.12). MCC particles can be clearly seen in the fracture surfaces, but the dispersion and the distribution is optimal. Although samples with US treatment contain micro sized particles (Fig. 5.12), which involves that the size reduction was not complete, other results indicate that the samples contain smaller particles as well.

neat PLA   PLA + 10 wt% PEG400

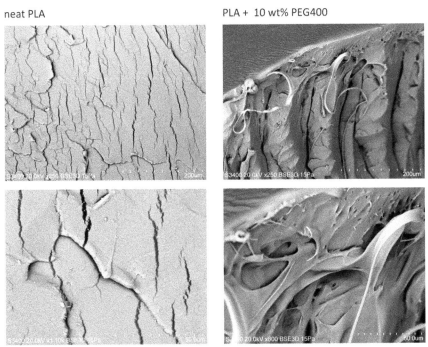

FIGURE 5.10 SEM micrographs of the tensile fractured surfaces of neat PLA and PLA+10 wt.% PEG samples.

PLA + 1 wt% MCC + 10 wt% PEG400   PLA + 3 wt% MCC + 10 wt% PEG400   PLA5 + 5 wt% MCC + 10 wt% PEG400

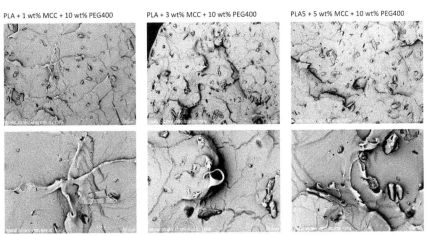

FIGURE 5.11 SEM micrographs of the tensile fractured surfaces of composites containing MCC and PEG400.

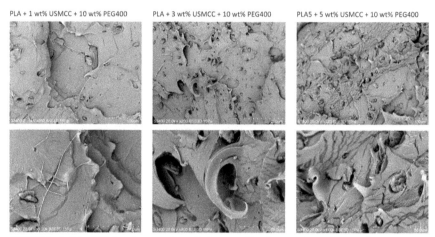

FIGURE 5.12 SEM micrographs of the tensile fractured surfaces of composites containing USMCC and PEG400.

5.4   CONCLUSIONS

During this research MCC, ultrasound treated MCC and PEG400 was used to modify the properties of the PLA. According to the results treatment of MCC in PEG 400 media led to the formation of spherical cellulose nanocrystals which have remarkable effect on the properties of the PLA matrix. Small amount (1 wt.%) of US treated cellulose in PEG 400 was enough to reduce remarkably the brittleness of the PLA and to create a though material. Slighter improvement occurred when PEG400 or MCC with PEG400 were only used. The thermal properties of the foils did not change significantly except the cold crystallization temperature, which shifted to lower values attributed to chain mobilizing effect of the plasticizer. The presence of PEG 400 and MCC and especially the US treated MCC indicated higher crystallinity. Although amorphous characteristic was stronger probably due to the fast cooling rates during the film extrusion process, WAXD results showed enhancement in the crystallinity of cellulose, which was attributed to the ultrasound treatment. While SEM images showed that micro sized cellulose particles still remained (despite the US treatment) TEM images proved that the size of the particles were reduced remarkably too. Possibly a hybrid of micro and nanocomposite was formed. Although further research is needed to improve the effectiveness of ultrasound treatment to achieve a homogeneous nanocrystal fraction, using US treated MCC in PEG media as a reinforcement can highly improve the toughness of the rigid and brittle poly (lactic acid) matrix. Varying the processing parameters can involve further modifying effects since, as Sanchez-Garcia and Lagaron [41] and Garner et al. [56] highlighted, the CNC processing method has a high impact on the further properties of the PLA-CNC nanocomposites.

KEYWORDS

- **Cellulose Nanocrystal**
- **Composite**
- **Poly (Lactic Acid)**

REFERENCES

1. Gajria, A. M., Dave, V., & Gross, R. A. (1996). Miscibility and Biodegradability of Blends of poly (lactic acid) and poly (vinyl acetate) Polymer, 37(3), 437–444.

2. Martin, O., & Avérous, L. (2001). Poly (lactic acid) Plasticization and Properties of Biodegradable Multiphase Systems Polymer, 42, 6209–6219.

3. Ljungberg, N., & Wesslén, B. (2002). The Effects of Plasticizers on the Dynamic Mechanical and Thermal Properties of poly (lactic acid), Journal of Applied Polymer Science, 86, 1227–1234.

4. Ljungberg, N., & Wesslén, B. (2003). Tributyl Citrate Oligomers as Plasticizers for poly (lactic acid), Thermo-Mechanical Film Properties and aging Polymer, 44, 7679–7688.

5. Baiardo, M., Frisoni, G., Scandola, M., Rimelen, M., Lips, D., Ruffieux, K., & Wintermantel, E. (2003). Thermal and Mechanical Properties of Plasticized poly (L-lactic acid) Journal of Applied Polymer Science, 90, 1731–1738.

6. Rasal, R. M., Janorkat, A. V., & Hirt, D. E. (2010). Poly (lactic acid) Modifications Progress in Polymer Science, 35, 338–356.

7. Balakrishnan, H. (2010). Mechanical, Thermal and Morphological Properties of Poly lactic acid/linear Low Density Polyethylene Blends, 42(3), 223–239.

8. Tuba, F., Oláh, L., & Nagy, P. (2011). Characterization of Reactively Compatibilized Poly (D, L-lactide)/Poly (e-caprolactone) Biodegradable Blends by Essential Work of Fracture Method Engineering Fracture Mechanics, 78, 3123–3133.

9. Takayama, T., Todo, M., & Tsuji, H. (2006). Improvement of impact fracture properties of PLA/PCL polymer blend due to LTI addition Journal of Material Sciences, 41, 4989–4992.

10. Takayama, T., Todo, M., & Tsuji, H. (2011). Effect of Annealing on the Mechanical Properties of PLA/PCL and PLA/PCL/LTI Polymer Blends, Journal of the Mechanical Behavior of Biomedical Materials, 4(3), 255–260.

11. Garlotta, D., Doane, W., Shogren, R., Lawton, J., & Willet, J. L. (2003). Mechanical and Thermal Properties of Starch-filled Poly (D, L-Lactic acid/poly (hydroxy ester ether) Biodegradabe Blends Journal of Applied Polymer Science, 88, 1775–1786.

12. Shogren, R. L., Doane, W. M., Garlotta, D., Lawton, J. W., & Willett, J. L. (2003). Biodegradation of Starch/polylactic acid/poly (hydroxyester-ether) Composite Bar Sin Soil Polymer Degradation and Stability, 79, 405–411.

13. Oksman, K., Skrifvars, M., & Selin, J. F. (2003). Natural Fibres as Reinforcement in Polylactic Acid (PLA) Composites Science and technology, 63, 1317–1324.

14. Islam, M. S., Pickering, K. L., & Foreman, N. J. (2010). Influence of Alkali Treatment on the Interfacial and Physicomechanical Properties of Industrial Hemp Fiber Reinforced Polylactic Acid Composites, Part A: Applied Science and Manufacturing, 41(5), 596–603.

15. Sawpan, A. M., Pickering, L. K., Fernyhough, A. (2011). Effect of Fiber Treatments on Interfacial Shear Strength of hemp fiber reinforced polylactide and unsaturated polyester composites, 42, 1189–1196.

16. Ray, S. S., Pralay, M., Okamoto, M., Yamada, K., & Ueda, K. (2002). New Polylactide/layeres Silicate Nanocomposites, 1. Preparation, Characterization, and Properties Macromolecules, 35, 3104–3110.

17. Ray, S. S., Yamada, K., Okamoto, M., Fujimoto, Y., Ogami, A., & Ueda, K. (2003). New Polylactide/layered Silicate nanocomposite, 5. Designing of Materials with Desired Properties Polymer, 44, 6633–6646.

18. Shibata, M., Someya, Y., Orihara, M., & Miyoshi, M. (2005). Thermal and Mechanical Properties of Plasticized Poly (L-lactide) nanocomposites with organo-modified montmorillonites, Journal of Applied Polymer Science, 99(5), 2594–2602

19. Jiang, L., Zhang, J., & Wolcott, M. P. (2007). Comparison of Polylactide/nano-sized Calcium Carbonate and Polylactide/montmorillonite Composites, Reinforcing Effects and Toughening Mechanisms Polymer, 48, 7632–7644.

20. Krishnamachari, P., Zhang, J., Yan, J., Shahbzi, A., Uitenham, L., & Lou, J. (2007). Thermal Characterization of Biodegradable Poly (lactic acid)/clay nanocomposites, In Proceedings of the (Nation Conference on Environmental Science and Technology, Houston, Texas, USA, August 6–9. 2007, Steven, K. S. Ed., American Science Press, Houston USA (2007), 219–225.

21. Gamez-Perez, J., Nascimento, L., Bou, J. J., FrancoUrquiza, E., Santana, O. O., Carracso, F., & Maspoch, M. L. (2011). Influence of Crystallinity on the Fracture Toughness of Poly (lactic acid)/montmorillonite Nanocomposite Prepared by Twin-Screw Extrusion, Journal of Applied Polymer Science, 12, 896–905.

22. Ozkoc, G., & Kemaloglu, S. (2009). Morphology, biodegradability, mechanical, and thermal properties of nanocomposite films based on PLA and plasticized PLA Journal of Applied Polymer Science, 144, 2481–2487.

23. Rodríguez-Llamazares, S., Rivas, L. B., Pérez, M., & Perrin-Sarazin, F. (2012). Poly(ethylene glycol) as a Compatibilizer and Plasticizer of Poly (lactic acid)/clay nanocomposite, High Performance Polymers, 24, 254–261.

24. Pluta, M., Jeszka, J. K., Boiteux, G. Polylactide/montmorillonite nanocomposites: Structure, dielectric, viscoelastic and thermal properties European Polymer Journal (2007), 43, 2819–2835.

25. Chiu, W. M., Chang, Y. A., Kuo, H. Y., Lin, M. H., & Wen, H. C. (2008). A Study of Carbon nanotubes/biodegradable Plastic Polylactic acid Composites Journal of Applied Polymer Sciences, 108, 3024–3030.

26. Balakrishnan, H., Hassan, A., Wahit, M. U., Yussuf, A. A., & Razak, S. B. (2010). A. Novel toughened polylactic acid nanocomposite, Mechanical, thermal and morphological properties Materials and Design, 31, 3289–3298.

27. Samir, M. A., Alloin, F., Sanchez, J. Y., & Dufresne, A. (2004). Cellulose nanocrystals reinforced poly (oxyethylene), Polymer, 45, 4149–4157.

28. John, M. J., & Thomas, S. (2008). Biofibers and Biocomposites Carbohydrate Polymers, 71(3), 343–364.

29. Mathew, A. P., Oksman, K., & Sain, M. (2005). Mechanical Properties of Biodegradable Composites from Polylactic acid (PLA) and Microcrystalline Cellulose (MCC), Journal of Applied Polymer Science, 97, 2014–2025.

30. Petterson, L., & Oksman, K. (2006). Biopolymer Based Nanocomposite, Comparing Layered Silicates and Microcrystalline Cellulose as Nano reinforcement Composite Science and Technology, 66, 2187–2196.

31. Suchaiya, V., & Aht-Ong, D. (2011). Effect of Microcrystalline Cellulose from Banana Stem Fiber on Mechanical Properties and Crystallinity of PLA Composite Films, Material Science Forum, 695, 170–173.

32. Iwatake, A., Nogi, M., & Yano, H. (2008). Cellulose Nanofiber-reinforced Polylactic acid Composites Science and Technology, 68, 2103–2106.

33. Nakagaito, A. N., Fujimura, A., Sakai, T., Hama, Y., & Yano, H. (2009). Production of Micro Fibrillated Cellulose (MFC) reinforced Poly lactic acid (PLA) nano composites from sheets obtained by a papermaking-like process Composites Science and Technology, 69, 1293–1297.

34. Qu, P., Gao, Y., Wu, G., & Zhang, L. (2010). Nanocomposites of Poly (lactic acid) Reinforcement with Cellulose nanofibrils BioResources, 5(3), 1881–1823.

35. Kowalczyk, M., Piorkowska, E., Kulpinksi, P., & Pracella, M. (2010). Mechanical and Thermal Properties of PLA Composites with Cellulose Nanofibers and Standard Size Fibers Composites, Part A, 42, 1509–1514.

36. Wang, B., & Mohini, S. (2007). The Effect of Chemically Coated Nanofiber Reinforcement on Biopolymer Based Nanocomposites BioResources, 02, 371–384.

37. Oksman, K., Matthew, A. P., Bondeson, D., & Kvien, I. (2006). Manufacturing Process of Cellulose Whiskers/polylactic acid nanocomposites Composites Science and Technology, 66(15), 2776–2784.

38. Bondeson, D., & Oksman, K. (2007). Dispersion and Characteristics of Surfactant Modified Cellulose Whiskers Nanocomposites Composite Interface, 14(7–9), 617–630.

39. Bondeson, D., & Oksman, K. (2007). Polylactic acid/cellulose Whisker nanocomposites Composites: Part A, 38, 2486–2492.

40. Petersson, L., Kvien, I., & Oksman, K. (2007). Structure and Thermal Properties of Poly (lactic acid)/Cellulose Whiskers Nanocomposite Materials Composite Science and Technology, 67, 2535–2544.

41. Sanchez, M. D., & Lagaron, J. M. (2010). On the use of Plant Cellulose Nanowhiskers to Enhance the Barrier Properties of polylactic acid Cellulose, 17, 987–1004.

42. Fortunati, E., Armentato, I., Zhou, Q., Iannoni, A., Visai, L., Berglund, L. A., & Kenny, J. M. (2012). Multifunctional Bionanocomposite films of Poly (lactic acid), Cellulose Nanocrystals and Silver Nanoparticles Carbohydrate Polymers, 87, 1596–1605.

43. Jacobsen, S., & Fritz, H. G. (1999). Plasticizing Polylactide the effect of different Plasticizers on the Mechanical Properties, Polymer Engineering and Science, 39(7), 1303–1310.

44. Baiardo, M., Frisoni, G., Scandola, M., Rimelen, M., Lips, D., Ruffieux, K., & Wintermantel, E. (2003). Thermal and Mechanical Properties of Plasticized Poly (L-lactic acid), Journal of Applied Polymer Science, 90, 1731–1738.

45. Cao N, Yang X, & Fu Y. (2009). Effects of Various Plasticizers on Mechanical and Water Vapor Barrier Properties of Gelatin Films, Food Hydrocolloids, 23, 729–735.

46. Lim, L. T., Auras, R., & Rubino, M. (2008). Processing Technologies for Poly (lactic acid) Progress in Polymer Science, 33, 820–852.

47. Furuhashi, Y., & Yoshie, N. (2012). Stereocomplexation of solvent-cast Poly (lactic acid) by Addition of Non-solvents, Polym Int., 60, 301–306.

48. Chen, X., Kalish, J., & Hu, S. L. (2011). Structure Evolution of α-phase Poly (lactic acid) Journal of Polymer Science, Part B Polymer Physics, 49, 1446–1454.

49. Filson, P. B., & Dawson Andoh, B. E. (2009). Sono-chemical Preparation of Cellulose Nanocrystals from Lignocellulose Derived Materials Bioresource Technology, 100, 2259–2264.

50. Yasuniwa, M., Tsubakihara, S., Sugimoto, Y., & Nakafuku, C. (2003). Thermal Analysis of the Double-Melting Behavior of Poly (L-lactic acid) Journal of Polymer Science, Part B: Polymer Physics, 42, 25–32.

51. Radjabian, M., Kish, M. H., & Mohammadi, N. (2010). Characterization of Poly (lactic acid) Multifilament Yarns. I. The Structure and Thermal Behavior Journal of Applied Polymer Science, 117, 1516–1525.

52. Pei, A., Zhou, Q., & Berglund, L. A. (2010). Functionalized Cellulose Nanocrystals as Biobased Nucleation Agents in Poly (L-lactide) (PLLA) Crystallization and Mechanical Property Effects Composites Science and Technology, 70, 815–821.

53. Ali, F., Chang, Y. W., Kang, S. C., & Yoon, J. Y. (2009). Thermal, Mechanical and Rheological Properties of Poly (lactic acid)/epoxidized Soybean Oil Blends Polymer Bulletin, 62, 91–98.

54. Gurun, B. (2010). Deformation Studies of Polymers and Polymer/clay nanocomposites, PhD Dissertation, Georgia Institute of Technology.

55. Cotterell, B., Chia, J. Y. H., & Habieb, K. (2007). Fracture Mechanisms and Fracture Toughness in Semicrystalline Polymer Nanocomposites Engineering Fracture Mechanics, 74(7), 1054–1078.

56. Gardner, D. J., Oporto, G. S., Mills, R., & Samir, M. A. (2008). S. A. Adhesion and Surface Issues in Cellulose and Nano cellulose, Journal of Adhesion Science and Technology, 22, 545–567.

CHAPTER 6

AN EXPERIMENTAL OBSERVATION OF DISPARITY IN MECHANICAL PROPERTIES OF TURMERIC FIBER REINFORCED POLYESTER COMPOSITES

NADENDLA SRINIVASABABU, SURESH KUMAR,
and K. VIJAYA KUMAR REDDY

CONTENTS

Abstract 124

6.1   Introduction 124

6.2   Experimental Work 125

6.3   Results and Discussion 126

6.4   Conclusions 133

Keywords 133

References 134

ABSTRACT

Utilization of eco friendly materials to suit the day-to-day needs invite the look for nature available materials like natural fibers. In the present work turmeric fiber is extracted from the stem and petiole by rolling method. The extracted fiber is chemically treated and is reinforced in the polyester matrix for making composites. The fabricated specimens are tested to determine tensile, specific tensile, flexural, specific flexural and impact strength properties as per ASTM procedures. Morphology of untreated and treated fibers is also examined using SEM.

6.1   INTRODUCTION

The reinforcement of natural fibers in various applications is increasing due to their superior specific properties. Many researchers have made several investigations on natural fiber reinforced polymer composites to see their performance under mechanical, thermal and electrical loadings. Out of which the sample of results is outlined below.

Jayamol George et al. [1] showed that the mechanical properties of pineapple leaf fiber reinforced LDPE composites were enhanced and elongation at break reduced with increase in fiber loading. A composite material consisting of HDPE, sand and short henequen fibers has been developed by Herrera-Franco et al. [2], and characterized for tensile and flexural properties. They also reported that the tensile strength of the HDPE-sand composites does not seem to be affected by the processing temperature, for any filler content, but the tensile modulus shows similar behavior for filler contents greater than 15% w/w.

The effect of silane coupling agent on curing characteristics and mechanical properties of bamboo fiber filled natural composites were studied by Hanafi Ismail et al. [3]. Plastic fiber composites consisting of polypropylene or polyethylene and pine wood, big blue stem, soybean hulls, or distillers dried grain and solubles were prepared by James et al. [4] for evaluating young's modulus, tensile, flexural strengths, and impact energy [5] reported the flexural and tensile strengths of the 30% agro fiber composites increased by more than 60% with Epolene G-3015.

Roger H. Newman et al. [6] treated leaf fibers from phromium tenax (New Zealand flax) with 1 wt.% aqueous NaOH at 30 °C to remove all of the acetyl groups, accounting for most of a 7% mass loss [7] studied the influence of silane degradation on the mechanical properties of vinyl ester composites reinforced with glass, sisal and coconut fibers and natural fibers modified with bitumen. Flexural and impact properties of turmeric fiber reinforced composites is determined by Srinivasababu et al. [8], and compared the results with other composites reinforced with various fibers.

Turmeric is known as the 'golden spice' and 'spice of life.' It has been used in India as a medicinal plant, and held sacred from time immemorial. In India the state of Andhra Pradesh is the largest producer of turmeric, with an area of 64,000 ha and a production of 3, 46, 000 t. Botanically turmeric is called "Curcuma Longa." Turmeric importance was lucidly described by Ravindran et al. [9].

Utilization of discarded turmeric stems for making composites and using them for various engineering applications is obviously a significant subject. In 2010, Srinivasababu et al. [10] reported the tensile properties of turmeric fiber reinforced polyester composites reinforced with chemically treated fiber at various concentrations.

6.2 EXPERIMENTAL WORK

6.2.1 EXTRACTION OF FIBER

In the present work turmeric stems are taken from the field of Chenna-kesavaiah, Gorigapudi, Guntur Dt. and are used to extract the fiber by Rolling method introduced by Srinivasababu [10] and is described below. Stems of turmeric are placed between two hard Bunwar rubber sheets. A roller of 40 mm diameter is allowed to roll manually on the top surface of the rubber sheet slowly. Then the split fibers are carefully detached from the bonding portions if any present, using a splitter of 0.5 mm diameter needle. The extracted fiber is segregated into turmeric stem fiber (TSF), turmeric petiole fiber (TPF).

6.2.2 CHEMICAL TREATMENT

Both the extracted fibers (turmeric stem and petiole) are chemically treated (CT) at ambient temperature. Initially turmeric stem fibers are treated with 0.125 M NaOH (45 min), 0.007909 M $KMnO_4$ (5 min) and 1.40701E-05

H_2SO_4 (5 min). From now onwards the treated fibers are called as turmeric stem CT (TSF-CT). Turmeric petiole fibers are treated with 0.125 M NaOH (45 min), 0.01266 M $KMnO_4$ (5 min) and 7.50408E-06 H_2SO_4 (5 min). The chemically treated turmeric petiole fibers are called as TPF-CT. In case of the flexural test composite specimens turmeric stems are undergone for two chemical treatments, that is, CT-1, CT-2. The treating procedure in CT-1 is, turmeric stems are soaked in 0.375 M NaOH for 10h 45 min, then 0.5062 $KMnO_4$ for 20 min and finally 3.75204E-06 M H_2SO_4 whereas in CT-2 turmeric fibers are only soaked in H_2SO_4 for 10 h 10 min at 0.1111 M concentration. All the chemically treated fibers are washed with huge quantity of distilled water and are dried at ambient conditions for 24 h and then they are placed in a heating oven at 70 °C for 2 h.

6.2.3 FABRICATION AND TESTING OF COMPOSITES

Untreated and chemically treated turmeric fiber is reinforced into Ecmalon 4413 polyester matrix for making turmeric fiber reinforced polyester composites by hand lay-up technique at room temperature. The fabricated composites are tested to determine the tensile [8], flexural and impact [10] properties as per ASTM procedures. Fiber morphology is studied using Jeol JSM-5350 Scanning Electron Microscope (SEM).

6.3 RESULTS AND DISCUSSION

This chapter concentrates on the observation of the SEM image of the untreated and chemically treated fiber and to find out possible reasons for improvement of the mechanical properties of the composites. Waxy materials are observed in turmeric petiole fiber from SEM micrograph (Fig. 6.1a), and after chemical treatment rougher surface and clear void is observed in Fig. 6.1b. Ribbon like structure with coated waxy and other materials on filaments is observed in case of turmeric stem fiber (Fig. 6.2a). Clear filaments having channels on its surface are observed in chemically treated fiber (Fig. 6.2b).

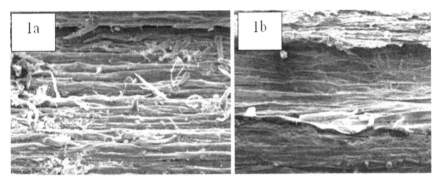

FIGURE 6.1   Turmeric Stem fiber before and after CT.

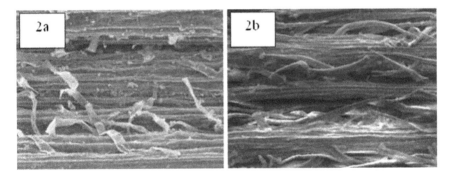

FIGURE 6.2   Turmeric Petiole fiber before and after CT.

The composites fabricated with untreated turmeric fibers are de-bonded from the matrix due to unlock between the fibers and matrix. This may be due to waxy nature of fiber and moisture present in it. Air bubbles inside the turmeric fiber reinforced composite specimen are visualized. This kind of trend is observed in all the 200 fabricated specimens. Hence all tensile test specimens are prepared with chemically treated turmeric fibers as reinforcement in the polyester matrix. Tensile tested turmeric fiber reinforced composites have shown disparity in their mechanical properties when reinforced with turmeric stem, petiole fibers. The specific properties of the composites are also determined.

The tensile strength of TSF reinforced polyester composites had shown increasing trend at all the volume fraction than that of TP FRP composites

(Fig. 6.3). Specific tensile strength of chemically treated turmeric stem fiber (TSF CT) reinforced polyester composites exhibited highest value of 27.36 MPa and is higher than other composites investigated in the chapter (Fig. 6.5).

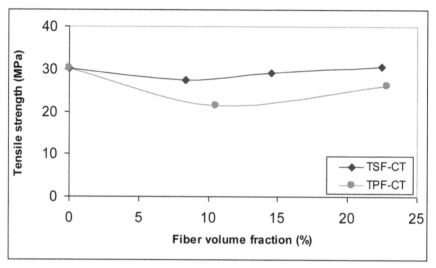

FIGURE 6.3   Variation of tensile strength with increase in fiber volume fraction.

Tensile modulus of TS FRP composites is 1.186 times higher than that of the TP FRP composites at maximum fiber volume fraction and is shown in Fig. 6.4. Chemically treated turmeric stem fiber (TSF CT) reinforced polyester composites shown 15.15% more specific tensile modulus than chemically treated turmeric petiole fiber (TPF CT) reinforced polyester composites at maximum fiber volume fraction of fiber, is observed from Fig. 6.6.

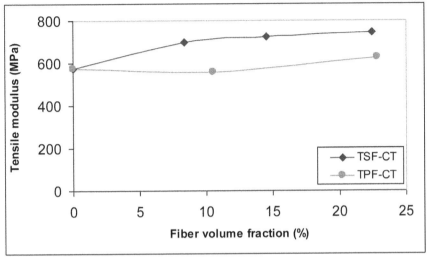

FIGURE 6.4   Variation of tensile modulus with increase in fiber volume fraction.

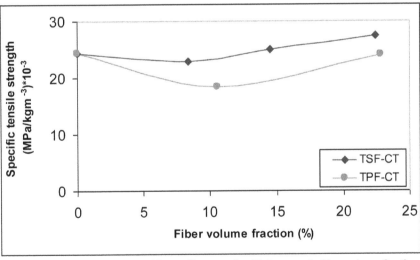

FIGURE 6.5   Variation of specific tensile strength with increase in fiber volume fraction.

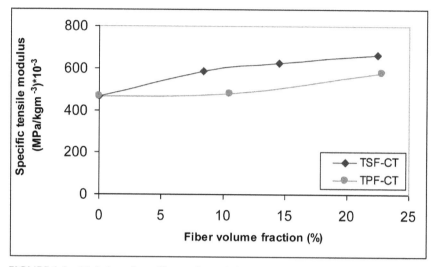

FIGURE 6.6   Variation of specific tensile modulus with increase in fiber volume fraction.

Flexural strength, modulus, specific flexural strength, modulus are graphically represented against fiber volume fraction in Figs. 6.7–6.10, respectively. With CT-2 of turmeric fibers flexural strength of turmeric FRP composites had shown increasing trend with increase in fiber content whereas in case of turmeric CT-1 FRP composites the maximum value of flexural strength 90.29 MPa is achieved at 21.01% V_f. Highest flexural modulus of 11.4 GPa is obtained for turmeric fibers CT-1 when compared with turmeric fibers CT-2 FRP composites. Similar kind of trend is observed in case of specific flexural strength and modulus properties of turmeric FRP composites. All the composite specimens have failed due to bending load only at the outer most layers.

Due to relatively more width of the specimen compared to depth node-bonding of the turmeric fibers from the matrix is identified. Hence untreated turmeric stem, petiole fiber reinforced composites are tested under impact loading and the impact strength is determined.

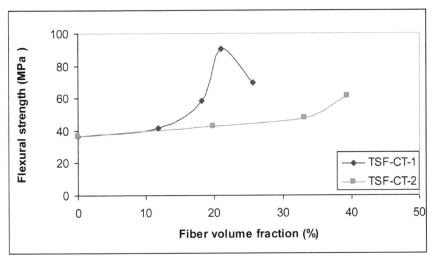

FIGURE 6.7   Variation of flexural strength with increase in fiber volume fraction.

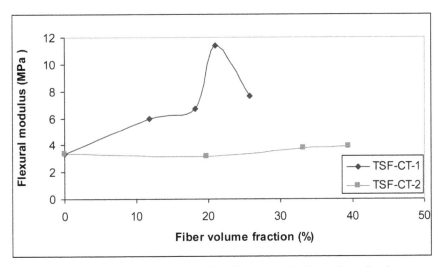

FIGURE 6.8   Variation of flexural modulus with increase in fiber volume fraction.

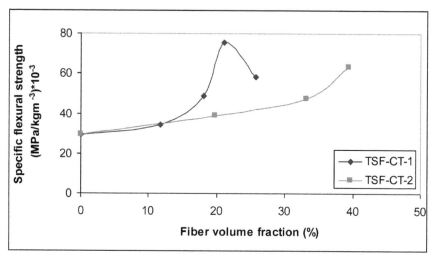

FIGURE 6.9   Variation of specific flexural strength with increase in fiber volume fraction.

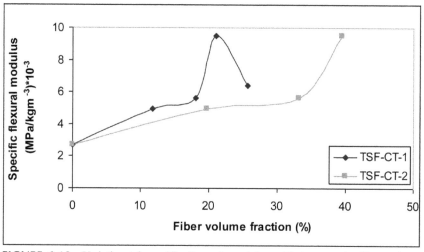

FIGURE 6.10   Variation of specific flexural modulus with increase in fiber volume fraction.

Impact strength against fiber volume fraction is graphically represented in Fig. 6.11 and 10.35 kJ/m² is recorded for turmeric petiole fiber reinforced polyester composites at maximum fiber volume fraction and is 1.87% more than turmeric stem fiber reinforced polyester composites.

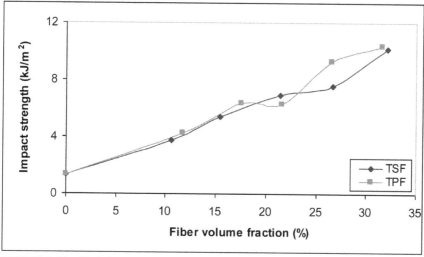

FIGURE 6.11   Variation of impact strength with increase in fiber volume fraction.

6.4   CONCLUSIONS

Chemically treated turmeric stem fiber reinforced polyester composites exhibited more specific tensile strength and modulus than the composites reinforced with chemically treated turmeric petiole fiber. Impact strength of turmeric petiole fiber composites is higher than stem fiber reinforced composites.

Utilization of agro waste, that is, turmeric stems for making composites, which are light in weight, and reasonably good strength is obviously a significant subject.

KEYWORDS

- **Agro waste**
- **Mechanical testing**
- **Polyester composites**
- **Reinforcement**
- **Turmeric**
- **Turmeric fiber**

REFERENCES

1. Jaymol George, S. S., Bhagawan, N., Prabhakaran, & Sabu Thomas (1995). "Short Pineapple Leaf Fiber Reinforced Low Density Polyethylene Composites." Journal of Applied Polymer Science 57, 853–855.
2. Herrera-Franco, P., Valadez-Gonzalez, A., & Cervantes-Uc, M. (1997). Development and Characterization of a HDPE-sand-natural Fiber Composite, Composites Part B: Engineering, 28 B, 331–343.
3. Hanafi Ismail, S., Shuhelmy, M. R., & Edyham (2002). "The Effects of a Silane Coupling Agent on Curing Characteristics and Mechanical Properties of Bamboo Fiber Filled Natural Rubber Composites," European Polymer Journal, 38, 39–47.
4. James, L., Julson, Gurram Subbarao, D. D., Stokke, Heath H., & Gieselman (2004). "Mechanical Properties of Biorenewable Fiber/plastic composites," Journal of Applied Polymer Science, 93, 2484–2493.
5. Keener, T. J., Stuart, R. K., & Brown, T. K. (2004), "Maleated Coupling Agents for Natural Fiber Composites," Composites part a: Applied Science and Manufacturing, 35, 357–362.
6. Roger, H., Newman Evamaria, C., Clauss James, E. P., Carpenter, & Armin Thumm (2007). "Epoxy Composites Reinforced with Deacetylated Phormium Tenax Leaf Fibers," Composites part a: Applied Science and Manufacturing, 38, 2164–2170.
7. Nicolai, F. N. P., Botaro, V. R., & Cunha Lins, V. F. (2008). "Effect of Saline Degradation on the Mechanical Properties of Vinyl Ester Matrix Composites Reinforced with Glass and Natural Fibers." Journal of Applied Polymer Science, 108, 2494–2502.
8. Srinivasababu, N., Murali Mohan Rao, K., & Suresh Kumar, J. (2009). "Turmeric FRP Composites: Experimental Determination of Flexural and Impact Properties," Indian Journal of Fiber and Textile Research, 4(7), 1323–1332.
9. Ravindran, P. N., Nirmal Babu, K., & Sivaraman, K. "Turmeric," CRC Press, 1.
10. Srinivasababu, N., Murali Mohan Rao, K., & Suresh Kumar, J. (2010) "Tensile Properties of Turmeric Fiber Reinforced Polyester Composites," Indian Journal of Fiber and Textile Research, 35, 324–329.

CHAPTER 7

WAVELENGTH DEPENDENCE OF SERS SPECTRA OF PYRENE

F. HUBENTHAL, D. BLÁZQUEZ SÁNCHEZ, R. OSSIG,
H. SCHMIDT, and H.-D. KRONFELDT

CONTENTS

Abstract 136
7.1   Introduction.. 136
7.2   Experimental... 137
7.3   Characterization 138
7.4   Results and Discussion 140
7.5   Conclusions.. 142
Keywords 142
References.. 143

ABSTRACT

In this contribution, we investigated the wavelength dependence of sur-face-enhanced Raman scattering (SERS) spectra of pyrene. For this pur-pose, three different excitation wavelengths have been applied: λ=514.5 nm, λ=647.0 nm and λ=785.0 nm. To achieve comparable experimental conditions, tailor-made SERS substrates were prepared, consisting of sil-ver or gold nanoparticles on sapphire supports. The tailoring is based on control of the mean shape and size of the nanoparticles, which permits a precise tuning of their localized surface plasmon polariton resonance in the vicinity of the excitation wavelength, used for the SERS measure-ments. We demonstrate that greatly enhanced SERS spectra were obtained for all wavelengths. However, only for the excitation wavelength of λ=785 nm, almost background free SERS spectra of pyrene with an excellent signal-to-noise ratio were obtained.

7.1   INTRODUCTION

The optical properties of noble metal nanoparticles (NPs) are dominated by a collective oscillation of conduction band electrons, that is, by localized surface plasmon polariton resonances. For simplicity, we refer to this kind of excitations as plasmon resonance or simply plasmon. The excitation of a plasmon is accompanied by a local electromagnetic field enhancement in the vicinity of the NP surface, which is exploited in numerous applications [1–13]. One prominent example is Raman scattering [14–23], which is a vibrational spectroscopic technique, where incident laser light is inelasti-cally scattered by molecules. Raman scattering allows substance identi-fication due to its inherent molecule fingerprinting capability. However, the Raman cross-section is extremely low. To overcome this drawback surface enhanced Raman spectroscopy (SERS) can be applied, where the molecules are attached to metallic nanostructures. The metal structure acts as an antenna, which does not only enhance the incident light field, but also the Raman scattered field. Thus, SERS has the potential to combine the sensitivity of fluorescence with the structural information of Raman spectroscopy. The SERS enhancement depends critically on the optical properties, that is, on the morphology of the nanostructure [24]. There-fore, fabrication methods that produce SERS substrates with tunable optical properties are desired, which generate an optimum enhancement of the Raman signal. To achieve such SERS substrates, for example, lithography techniques [4, 15, 25] and tailoring of supported metal NPs [26, 27] have been successfully applied.

In this contribution, we exploit the growth kinetics of metal NPs on sapphire supports, to generate tailored NP ensembles. By controlling the shape and size of the NPs during growth, we have tuned the plasmon resonance in the vicinity of the SERS excitation wavelengths. This allows measurements of SERS spectra as a function of the excitation wavelength under comparable experimental conditions.

7.2   EXPERIMENTAL

The NPs have been prepared by Volmer-Weber growth on sapphire supports in an ultra high vacuum set up. A heating system at the sample holder allows a controlled increase of the sample temperature up to T=500 K. A Xe-arc lamp (Osram, XBO 450 W/1) combined with a monochromator (AMKO, 1200 lines/mm, blaze 250 nm) were used to measure the optical spectra in situ, using p-polarized light with an angle of incidence of 45° with respect to the surface normal of the support. In addition, the NPs were characterized by AFM (ThermoMicroscopes, Autoprobe CP) under ambient conditions in noncontact mode. Details of the experimental set up, the NP preparation and characterization can be found elsewhere [26].

A micro Raman set up was employed to record the SERS spectra and three different wavelengths (514.5 nm, 647.0 nm, and 785.0 nm) were applied for SERS excitation. The backscattered light was collected by a microscope objective (Zeiss, F100), filtered with a notch plus filter (Kaiser) and analyzed by a spectrograph (Chromex 250 IS), equipped with a liquid-nitrogen cooled deep depletion CCD (Princeton Instruments, 400 × 1340 pixel). The set up had a spectral bandpass of 3 cm^{-1}. For the measurements, 2 µL of a 2 mM solution of pyrene (Fluka, purity >99%) in methanol (Merck, purity >99.5%) were spotted on the SERS substrates. After drying in air, the SERS response was optimized by adjusting the position of the substrate along the focal axis by a translation stage. For each wavelength several SERS spectra were measured with an integration time of 1 second and 3 accumulations each.

7.3 CHARACTERIZATION

The solid lines in Fig. 7.1a are extinction spectra of undisturbed grown gold NPs, for coverages ranging from Θ=0.8 10^{16} atoms/cm² to Θ=7.5 10^{16} atoms/cm². After deposition of Θ=7.5 10^{16} atoms/cm² the generated NPs have a mean equivalent radius of $<R_{eq}> = (18 \pm 3)$ nm and a mean axial ratio of $<a/b>$=0.07. Due to the growth kinetics, the axial ratio of the NPs drops off as the NPs increase in size and their plasmon resonance is shifted to lower photon energies. Since the NPs grow with an oblate shape [26, 28, 29], a thermal treatment as well as laser tailoring during growth [26, 30–32] allows a shaping of the NPs towards more spherical. Hence, both processes permit a tuning of the plasmon resonance to higher photon energies. An extinction spectrum of a laser tailored gold NP ensemble is depicted in Fig. 7.1a dashed line. Although the coverage was also Θ = 7.5 10^{16} atoms/cm², the mean axial ratio amounts to $<a/b>$ = 0.15 and coincides in the presented case almost with the excitation wavelength of λ = 785 nm. We emphasize, that the extinction amplitude of the tailored NP ensemble is slightly lower than for the naturally grown NPs with the same coverage. This can be explained by two effects. First, the position of the plasmon resonance is located at higher photon energies, where the dielectric function is different. Second, due to the heating of the NPs, evaporation of atoms may occur, which reduces the extinction of the NP ensemble.

Figure 7.1b displays an AFM image of gold NPs on a sapphire substrate. The coverage was Θ = 1.1 10^{16} atoms/cm². The NPs appear extremely broad and being in contact with each other. However, this is an artifact, due to the convolution with the probe tip. Detailed TEM investigations (not shown) revealed that the NPs are well separated.

Several series of SERS measurements have been accomplished for the three different excitation wavelengths, using pyrene as a probe molecule. Tailored SERS substrates (samples **1–3**) consisting of silver or gold NPs have been prepared on sapphire supports. The preparation parameters and the characteristics of the SERS substrates are summarized in Table 7.1.

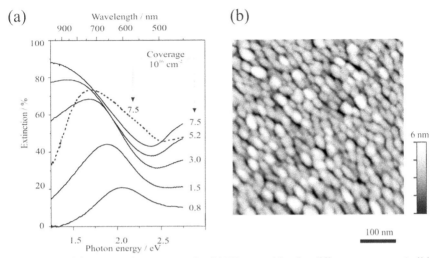

FIGURE 7.1(A) Extinction spectra of gold NP ensembles for different coverage (solid lines) and of a tailored NP ensemble (dashed line) for a coverage of $Q = 7.5 \cdot 10^{16}$ atoms/cm². (b) AFM image of undisturbed grown gold NPs.

TABLE 7.1 Preparation Parameters and Characteristics of the SERS substrates λ_{SPR} is the Central Wavelength of the Plasmon Resonance

Sample	1	2	3
material	silver	gold	gold
Θ / 10^{16} atoms/cm²	6.9 ± 0.9	7.6 ± 0.8	8.0 ± 1.0
$T_{support}$ / 10^2 K	3.3 ± 0.2	3.0 ± 0.2	3.0 ± 0.2
$<a/b>$	0.16	0.15	0.07
λ_{SPR} / nm	≈ 556	655	785

In all experiments, the plasmon resonance has been tuned approximately in coincidence with the excitation wavelength of the laser light for the SERS measurements. For the silver NPs (sample **1**), the tuning has been achieved by a moderate increase of the substrate temperature to T=330 K during NP growth. The elevated temperatures stimulate an increased diffusion of surface atoms. Hence, the NPs tend to grow in a more thermodynamical equilibrium and exhibit higher axial ratios compared to NPs prepared at lower temperatures. The tuning of the gold NPs (samples

2 and 3) has been achieved by laser tailoring, a method based on light induced heating of the NPs. To strengthen the manuscript and since laser tailoring is well described in the literature [16, 26, 32, 33] an explanation of the technique is omitted here. To assure comparable experimental conditions, the extinction maximum of the plasmon resonance is for all samples in the same order and between 75% and 85%.

7.4 RESULTS AND DISCUSSION

A Raman spectrum of pure pyrene is depicted in Fig. 7.2a, for comparison. The spectrum shows, at which Raman shifts between 500 cm^{-1} and 1900 cm^{-1} Raman lines occur. The inset in Fig. 7.2a displays schematically the structure of pyrene.

In the first set of SERS measurements, an excitation wavelength of $\square$=514.5 nm, generated by an argon ion laser, has been applied. Silver NPs were used, because for this wavelength they exhibit better optical properties compared to gold NPs. The SERS spectrum (Fig. 7.2b) shows all relevant peaks of pyrene, with an adequate signal-to-noise ratio (SNR). However, the spectrum has a significant background. In particular, the broad peaks at about 1300 cm^{-1} and 1550 cm^{-1} are caused by graphite formation due to the decomposition of the probe molecule. In addition, fluorescence may also contribute to the background. The background signal is a typical drawback for excitations with relatively high energetic laser light.

To avoid graphite formation and fluorescence, two additional sets of measurements have been performed, using λ=647 nm and λ=785 nm, generated by a krypton ion laser and a diode laser, respectively. Since for longer excitation wavelengths gold exhibits better optical properties compared to silver, gold NPs served as SERS substrates. In the SERS spectrum measured with an excitation wavelength of λ=647 nm (Fig. 7.2c), all relevant SERS lines are observed, except the line at 1065 cm^{-1}. This line is superimposed by two sapphire peaks at about 1050 cm^{-1}, which dominate the spectrum. The SNR is clearly worse compared to the spectrum obtained with an excitation wavelength of λ = 514.5 nm (cf. Fig. 7.2b) and a broad background is observed. The latter effect can be attributed again to decomposition of pyrene and graphite formation as well as to fluorescence. The low SNR results from a relatively weak field enhancement. At a wavelength of λ=647 nm, the plasmon resonance of gold NPs is strongly damped by the interband transition, which limits the field enhancement. Hence, the peaks are not so strong and the SNR is worse compared to the spectrum displayed in Figure 7.2b.

The situation changes drastically, if an excitation wavelength of λ=785 nm is used. In this case the SERS spectrum is almost background free (Fig. 7.2d). Hence, no fluorescence occurs and graphite formation is avoided, that is, pyrene does not decompose at this wavelength. Most importantly, the SNR is significantly better than in the previous spectra. It is even better than for pure pyrene (cf. Fig. 7.2a). Since no background signal occurs, an excellent SNR is obtained. In addition, the field enhancement is significantly better than for an excitation wavelength of λ=647 nm, because the plasmon resonance suffers no damping due to the interband transition. The measurements demonstrate how important the molecule/excitation wavelength combination is for an optimized SERS measurement. In particular, if molecules decompose under high energetic laser light, or if strong fluorescence appear measurements in the red or infrared region may deliver better results. Basic requirement are SERS substrates, which exhibit a plasmon resonance in this spectral region and posses high surface areas for the attachment of a sufficient number of molecules. These requirements are ideally fulfilled by the tailored gold NP ensembles, used in this study.

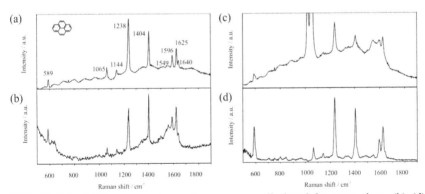

FIGURE 7.2 (a) Raman spectrum of pure pyrene; displayed for comparison. (b)–(d) SERS spectra of pyrene, obtained with excitation wavelengths of λ=514.5 nm (b), 647 nm (c), and 785 nm (d), respectively.

Note, that independently from the excitation wavelength, all NP ensembles are good amplifiers, generating SERS enhancement factors in the order of 10^5 to 10^6, as already discussed in a previous paper [16]. In the same paper, the effect of the tailoring on the SERS spectra has been discussed in detail. The influence of the tailoring has been neglected in the present study, because no absolute enhancement factors have been compared.

7.5 CONCLUSIONS

In this contribution, SERS spectra of pyrene have been investigated as a function of the excitation wavelength. Tailored silver and gold NP ensembles have been prepared, whose plasmon resonance coincide with the SERS excitation wavelength. We demonstrated that the excitation wavelength has a significant influence on the SERS measurements. In particular, applying an excitation wavelength of λ=785 nm avoided a decomposition of the probe molecule as well as fluorescence. As a result, an almost background free SERS signal with an excellent SNR was obtained. The SNR was significantly better than the SNR of the SERS spectra measured with excitation wavelengths of λ=514.5 nm or λ=647 nm. Hence, for molecules that decompose under green or yellow light, or samples that exhibit strong fluorescence, excitation wavelengths in the near infrared might be chosen. Along these lines, we have shown that gold NPs are ideal SERS substrates, which yield almost background free and strongly enhanced SERS spectra. Basic requirement for the experiments are SERS substrates with easily tunable optical properties, as presented in this study.

KEYWORDS

- Gold nanoparticles
- Optical properties
- Plasmon resonance
- Pyrene
- SERS
- SNR

REFERENCES

1. Bek, A., Jansen, R., Ringler, M., Mayilo, S., Klar, T., & Feldmann, J. (2008). Nano Lett., 8, 485.
2. Constantino, C., & Aroca, R. J. (2000). Ram. Spec., 31, 887.
3. Lakowicz, J., Shen, B., Gryczynski, Z., D'Auria, S., & Gryczynski, I. (2001). Biochem. and Biophys. Res. Com., 286, 875.
4. Felidj, N., Truong, S., Aubard, J., Levi, G., Krenn, J., Hohenau, A., Leitner, A., & Aussenegg, F. J. (2004). Chem. Phys., 120, 7141.
5. Gryczynski, I., Malicka, J., Shen, Y., Gryczynski, Z., & Lakowicz, J. (2002). J. Phys. Chem. B, 106, 2191.
6. Priebe, A., Sinther, M., Fahsold, G., & Pucci, A. J. (2003). Chem. Phys., 119, 4887.
7. Li, K., Stockman, M., & Bergman, D. (2005). Phys. Rev. B, 72, 153401.
8. Alschinger, M., Maniak, M., Stietz, F., Vartanyan, T., & Träger, F. (2003). Appl. Phys. B, 76, 771.
9. Rycenga, M., Cobley, C. M., Zeng, J., Li, W., Moran, C. H., Zhang, Q., Qin, D., & Xia, Y. (2011). Chem. Rev., 111, 3669.
10. Sancho-Parramon, J., Janicki, V., Loncaric, M., Zorc, H., Dubcek, P., & Bernstorff, S. (2011). Appl. Phys. A, 103, 745.
11. Sadeghi, S. M. (2010). Nanotechnology, 21, 455401.
12. Hubenthal, F., Morarescu, R., Englert, L., Haag, L., Baumert, T., & Träger, F. (2009). Appl. Phys. Lett., 95, 063101.
13. Morarescu, R., Englert, L., Kolaric, B., Damman, P., Vallée, R. A. L., Baumert, T., & Hubenthal, F. (2011). Träger, F. J. Mater. Chem., 21, 4076.
14. Moskovits, M. J. Ram. Spect. 36, 485.
15. Sowoidnich, K., Schmidt, H., Maiwald, M., Sumpf, B., & Kronfeldt, H. D. *(2010)*. *Food Bioprocess Technol., 3, 878.*
16. Hubenthal, F., Blázquez Sánchez, D., Borg, N., Schmidt, H., Kronfeldt, H. D., & Träger, (2009). F. Appl. Phys. B, 95, 351.
17. Murphy, T., Lucht, S., Schmidt, H., & Kronfeldt, H. D. (2000). J. Raman Spectrosc., 31, 943.
18. Kneipp, K., Kneipp, H., Itzkan, I., Dasari, R., & Feld, M. (2002). J. Phys. Cond. Matt., 14, R597.
19. Nie, S., & Emroy, S. (1997). Science, 275, 1102.
20. Schmidt, H., Sowoidnich, H. K., & Kronfeldt, H. D. *(2010). Appl. Spectrosc. 64, 888.*
21. Wang, Y., Asefa, T., & Langmuir (2010). 26, 7469.
22. Bell, S. E. J., & McCourt, M. (2009). R. Phys. Chem. Chem. Phys., 11, 7455.
23. Speed, J. D., Johnson, R. P., Hugall, J. T., Lal, N. N., Bartlett, P. N., Baumberg, J. J., & Russell, A. E. (2011). Chem. Comm. 47, 6335.
24. Shalaev, V. (2000). Nonlinear Optics of Random Media, Springer, Berlin, Heidelberg.
25. Felidj, N., Aubard, J., Levi, G., Krenn, J., Salerno, M., Schieder, G., Lamprecht, B., Leitner, A., & Aussenegg, F. (2002). Phys. Rev. B, 65, 075419.
26. Hubenthal, F. (2009). Eur. J. Phys., 30, S49.
27. Blázquez Sánchez, D., Gallasch, L., Schmidt, H., Kronfeldt, H. D., Borg, N., Hubenthal, F., & Träger, F. (2006). Proc. SPIE, 6099, 609904.

28. Hubenthal, F., Borg, N., & Träger, F. (2008). Appl. Phys. B, 93, 39.
29. Hubenthal, F. (2011). Noble Metal Nanoparticles, Synthesis and Optical Properties, In: Comprehensive Nanoscience and Technology, Andrews, D. L., Scholes, G. D., & Wiederrecht, G. P. (eds.), Oxford: Academic Press, volume 1, 375–435.
30. Morarescu, R., Blázquez Sánchez, D., Borg, N., Vartanyan, T., Träger, F., & Hubenthal, F. (2009). Appl. Surf. Sci., 225, 9822.
31. Pyatenko, A., Yamaguchi, M., & Suzuki, M. (2009). J. Phys. Chem. C., 113, 9078.
32. Ouacha, H., Hendrich, C., Hubenthal, F., & Träger, F. (2005). Appl. Phys. B, 81, 663.
33. Hubenthal, F., Hendrich, C., Ouacha, H., Blázquez Sánchez, D., & Träger, F. (2005). Int. J. Mod. Phys. B, 19, 2604.

CHAPTER 8

EMERGING THERAPEUTIC APPLICATIONS OF BACTERIAL EXOPOLYSACCHARIDES

P. PRIYANKA, A. B. ARUN, and P. D. REKHA

CONTENTS

8.1   Introduction 146
8.2   Prospective Therapeutic Applications of Bacterial EPS 147
8.3   Production of EPS 158
8.4   Conclusions 159
Keywords 160
References 160

8.1 INTRODUCTION

Bacterial exopolysaccharides (EPS) are high molecular weight (10–1000 kDa) carbohydrate polymers produced by bacteria as secondary metabolites [1]. Exopolysaccharides have critical physiological function in bacterial cells ranging from cell adhesion to imparting adaptive response to environmental adversities such as, extreme temperature, pH, salinity etc. [2]. Most bacterial EPS are synthesized intra cellularly and exported to the extracellular environment as macromolecules. But, a few polysaccharides such as, levans and dextrans are synthesized and polymerized outside the cells by the action of secreted enzymes that convert the substrate into the polymer in the extracellular environment [3]. Secreted EPS separated from the bacterial cell serves as an intact biopolymer that can be processed and leveraged for various commercial applications.

Structural and chemical diversity of the EPS is expressed in terms of its monosaccharide composition, spatial arrangement of the monosaccharides, stereochemistry of the functional groups, molecular weight and the net charge. Most of the monosaccharides present in EPS are made up of pentoses (D-arabinose, D-ribose, D-xylose), hexoses (D-glucose, D-galactose, D-mannose, D-allose, L-rhamnose, L-fucose), amino sugars (D-glucosamine and D-galactosamine), acidic sugars (glucuronic acid and mannuronic acid) [4]. Rare monosaccharides like 4-amino-4, 6-dideoxy-2-O-methylmannose and bacillosamine (2, 4-diamino-2, 4, 6-trideoxy-D-glucose) are limited to a few species of *Vibrio* and *Bacillus,* respectively [5, 6]. Exopolysaccharides can be either homopolysaccharides or heteropolysaccharides. Homopolysaccharides are polymers made up of a single type of monosaccharide unit (e.g., dextran-a polymer of glucose) and heteropolysaccharides are composed of more than one type of monosaccharide as repeating unit (e.g., alginate-a polymer of guluronic acid and mannuronic acid). Polysaccharides are formed by the glycosidic linkages between the monosaccharide units. Exopolysaccharides formed by β-1, 4 or β-1, 3 linkages between the monosaccharide units are characterized to be rigid (e.g., xanthan) and EPS formed by α-1, 2 or α-1, 6 linked monosaccharide units are likely to be more flexible (e.g., dextran) [7]. These linkages build the polymeric structure of the EPS resulting in the formation of high molecular weight complexes [1].

Possible application of the EPS is determined by the functional groups present in it. The nature and degree of functionalization impart valuable properties to the EPS. For example in succinoglycan (an EPS produced by *Rhizobium meliloti*), substitution of succinyl groups (–CO–CH$_2$–CH$_2$–CO–) lowers the thermal stability and pseudoplasticity of the polymer [8]. Increased degree of acetylation (–COCH$_3$–) in xanthan gum (EPS produced by *Xanthomonas campestris*) leads to decreased viscosity, conversely higher degree of pyruvate (–CH$_3$COCOO–) substitution leads to increased viscosity [9]. The percentage substitution of uronic acid influences the ion exchange and gel strength properties of alginate [10]. Presence of functional groups like sulfate and phosphate enhance the therapeutic potential of EPS produced by *Alteromonas infernus* [11] and *Lactococcus lactis* [12], respectively.

Chemical composition of the EPS determines the thermal stability, rheological, gelling and other functional attributes and hence applications of the polymer. Understanding the chemistry of EPS gives an orientation towards its appropriate applications and by inducing chemical modifications EPS can also be tailored to suit specific application.

8.2 PROSPECTIVE THERAPEUTIC APPLICATIONS OF BACTERIAL EPS

There has long been a growing demand for biopolymers in pharmaceutical industry due to the biocompatibility and safety over synthetic alternatives. Biopolymers may be either produced by biological systems (e.g., polysaccharide) or may be synthesized chemically but are derived from biological starting materials (e.g., poly lactic acid). Polysaccharides produced by plants, macroalgae and bacteria, have been widely exploited for food and pharmaceutical applications. Microbial EPS have advantage over the plant or macroalgal EPS due to the ease in production at industrial scale, unaffected by seasonal variations, more rapid growth and hence product recovery, moreover they embody a stable supply and cost [13]. Improved understanding on the structural diversity of the EPS has attracted the researchers to investigate the functional properties of these polymers. As the diversity of bacteria is being explored, the discovery of diverse biopolymers such as polysaccharides is also emerging.

The inherent biocompatibility and apparent nontoxic nature of some bacterial EPS have prompted their use in various biotechnological applications [14]. The antioxidative, antithrombotic, antiangiogenic, antimetastatic, immunoproliferative, antiviral and wound healing properties find various therapeutic applications in human health and medicine [15–18]. A few EPS are also used for tissue engineering applications as they can form three-dimensional structures with uniform pore size [19]. The anti-aging, moisturizing and skin protection abilities of the EPS find use in cosmeceutics [20]. High viscosifying, emulsifying and gelling activities find applications in the food industrial sector and drug delivery systems [21–23]. Some EPS are also identified for heavy metal sequestration [24], surfactant like properties [25], which could possibly be employed for bioremediation and detoxification. Such versatile application demonstrated by EPS is due to the structural diversity of EPS, which is genetically controlled and specific to the species of bacteria producing it. Many types of EPS with structural diversity have been isolated and are well documented in the literature. However, only a few have been commercialized with relevant industrial and therapeutic applications (Table 8.1).

TABLE 8.1   Representative Strains of the Bacteria Producing Important EPS, Chemical Composition of the EPS and its Therapeutic Application

Bacteria	EPS and its chemical composition	Application
Gluconacetobacter xylinus [26]	Cellulose Linear homopolymer of β-1, 4 D-glucose.	Wound healing [27].
Leuconostoc mesenteroides [28]	Dextran Branched homopolymer of glucose linked by linear α-1,6 glycosidic bond and branched α-1, 3 linkage.	Drug delivery [29].
Sphingomonas elodea [30]	Gellan Linear anionic heteropolymer of D-glucuronic acid, L-rhamnose and two molecules of D-glucose.	Intervertebral disc regeneration [31].
Alcaligenes fecalis [32]	Curdlan Linear homopolysaccharide of β-1,3-linked glucose.	Drug delivery [33].
Azotobacter vinelandii [34]	Alginate Linear heteropolymer of β-1–4-D-mannuronic acid and α-L-guluronic acid.	Wound dressing [35].

TABLE 8.1   *(Continued)*

Bacteria	EPS and its chemical composition	Application
Spingomonas ATCC 31555ᵀ [36]	Wellan Anionic heteropolysaccharide of β-1,4 linked D-glucose, D-glucuronic acid and a side chain of L-rhamnose linked by β-1,3 linkage.	Stabilizer for cosmetic creams [37].
Xanthomonas campestris [38]	Xanthan Branched anionic heteropolysaccharide of α- D-mannose with an acetyl group, β-D-glucuronic acid and a terminal β-D-mannose unit linked to a pyruvate group.	Ophthalmic drug delivery [39].
Rhizobium meliloti [8]	Succinoglycan Anionic heteropolymer of D-glucose and D-galactose bonded by β-1, 3-, 1, 4- and 1, 6-linkage, with functional succinyl, acetyl and ketal groups.	Thickener for cosmetics [40].

8.2.1   EXOPOLYSACCHARIDES AS ANTIOXIDANT AGENTS

The presence of chelating functional groups like sulfates and phosphates along with (-OH) groups contained in the sugar backbone of the EPS enhance the redox potential of the EPS. It is known that transition metal ions like Fe^{2+} and Cu^+ are the key factors for the initiation of Fenton's reaction leading to the generation of potentially dangerous hydroxyl free radical in biological systems. The chelation of these transition metal ions by the functional groups on the polysaccharide chain inhibit the generation of hydroxyl radical or render the ions inactive or poorly active in the Fenton's reaction [41]. Hydroxyl radicals are known to cause DNA damage by attacking the –H atoms on the sugar backbone of DNA [42]. Hence, sugar backbone of the polysaccharide acts possibly as a competitive substrate for hydroxyl radical attack thus sparing vital cellular components from peroxidation and further damage.

Several EPS isolated from bacteria are reported to have antioxidant activities (Table 8.2). Exopolysaccharide produced by probiotic strains of bacteria *Lactobacillus plantarum* [43], *Bacillus coagulans* [44], *Strepto-*

coccus phocae P180[T] [45] show potent antioxidant activities against reactive oxygen species like hydroxyl radical, superoxide radical, hydrogen peroxide and inhibit lipid peroxidation in vitro. These studies project the prospective health beneficiary role of the EPS taken indirectly as probiotics.

Bacterium *Edwardsiella tarda* produces two types of mannans with different molecular weight, of these the lower molecular weight EPS shows higher radical scavenging activities in vitro [46].

TABLE 8.2 Chemical Composition of Some Bacterial EPS having Antioxidant Properties

Bacteria	Chemical composition of the EPS
Lactobacillus plantarum	Cell bound heteropolymer of glucose, galactose and galactosamine with traces of phosphate, molecular weight (MW) 8.5 ´ 10⁵ Da
	Heteropolymer of glucose, galactose and rhamnose with traces of phosphate MW 4 ´ 10⁴ Da [47]
Bacillus coagulans	Heteropolymer of galactose, mannose, fucose, glucose and glucosamine [48]
Streptococcus phocae P180[T]	Heteropolymer of arabinose, fructose and galactose, MW 2.8 ´ 10⁵ Da [45].
Edwardsiella tarda	Produces two EPS fractions MW 2.9 ´ 10⁴ Da and 7.0 ´ 10⁴ Da. Both are homopolymers of mannose linked by a linear chain of 1, 3 linkage and branched at 1, 2 position [46]

It is evident that accumulation of free radicals are key players in the pathogenesis of many degenerative diseases like atherosclerosis, cancer, inflammatory joint disease, asthma, diabetes, senile dementia, degenerative eye, etc. [49]. Hence, consumption of some highly antioxidant EPS as dietary supplements can be useful in prolonging good health.

8.2.2 EXOPOLYSACCHARIDES AS IMMUNOMODULATORY AGENTS

Immunomodulatory activities associated with some probiotic strains of lactic acid bacteria (LAB) owe to the EPS produced by them (Table 8.3).

The immune modulatory effects are brought about by the interaction of the immune cells and EPS component of the LAB. Exopolysaccharides stimulate the antigen presenting dendritic cells through toll-like receptors and leads to the augmentation of cellular immunity.

Exopolysaccharide produced by a bacterial strain (MK1) shows immunomodulatory effects by macrophage activation, increased phagocytic index, secretion of TNF-α (tumor nechrotic factor), IL-1β (interleukin) and nitric oxide release in mouse leukemia monocyte macrophage cell lines (RAW-264.7) [50].

The EPS produced by *Lactococcus lactis* ssp. *cremoris* KVS20^T is active as a B-cell mitogen in spleen cells isolated from C3H/HeJ mice [51]. Several other strains of *Lactococcus lactis* stimulate the production of IL's and tumor necrotic factors (TNF) in vitro in macrophage cell lines J774.1 [52]. EPS produced by *Lactobacillus rhamnosus* RW-9595 M^T [53], *Lactobacillus delbrueckii* ssp. *bulgaricus* [12] and a thermotolerant strain of *Bacillus licheniformis* [54] possesses immunomodulatory effects on immunocompetent cells in vitro. The immunomodulatory activity of the EPS produced by *Lactobacillus delbrueckii* ssp. *bulgaricus* is due to the presence of functional phosphate group in the EPS [55]. The EPS produced by *Enterobacter cloacae* Z0206^T enriched by the addition of selenium has shown enhanced immunomodulatory activities in immuno compromised mice [15] however, the exact mode of action of the EPS in immunomodulation needs to be further explored so as to establish a stronger validation of the activity.

TABLE 8.3 Chemical Composition of Some Bacterial EPS having Immunomodulatory Activities

Bacteria	Chemical composition of the EPS
MK1	Heteropolymer of glucose, rhamnose, galactose and galactosamine MW ranging between 1.06 ´ 10^4 to 5.5 ´ 10^4 Da [50].
Lactobacillus delbrueckii ssp. *bulgaricus* 1073R-1^T	Phosphopolysaccharide of glucose and galactose [12].
Lactococcus lactis ssp. *cremoris* KVS20^T	Phosphopolysaccharide of rhamnose, glucose and galactose [51].
Lactobacillus rhamnosus RW-9595 M^T	Heteropolymer of glucose, galactose and rhamnose with functional acetyl groups [56].

TABLE 8.3 *(Continued)*

Bacteria	Chemical composition of the EPS
Bacillus licheniformis DSM465[T]	Heteropolymer of mannose and glucose [57].
Enterobater cloacae Z0206[T]	Heteropolymer of glucose, galactose and mannose, MW 2.93×10^4 Da [15].

8.2.3 EXOPOLYSACCHARIDES AS AGENTS FOR CANCER THERAPY

Exopolysaccharides demonstrate anticancer activities, by either inducing apoptosis of carcinogenic cells, inhibiting metastasis or angiogenesis by acting as matrix metalloprotease inhibitors, etc. [58]. Structurally diverse forms of EPS produced by different bacteria have shown anticancer activity (Table 8.4), however, the exact chemistry and molecular interaction behind the activities still needs in depth studies.

Cell bound EPS produced by *Lactobacillus acidophilus* induces autophagic cell death in colon cancer cell lines by induction of autophagy proteins Beclin-1 and GRP78, as well as indirectly through the induction of antiapoptotic Bcl-2 and Pro-apoptotic Bak genes [59]. The EPS produced by *Bifidibacterium bifidum* inhibits proliferation of human gastric cancer cell line by decreasing the expression of cell telomerase rate-limiting factor hTERT mRNA [60].

β-glucan type EPS produced by *Rhizobium* sp N613[T] shows potent antitumor activities in sarcoma 180, hepatoma 22, and ehrlich ascites carcinoma tumor bearing mice model [61]. The EPS produced by a bacterium *Halomonas stenophila* following sulfation shows selective apoptotic activity on human T-cell leukemic cells. Sulfation of the native EPS up to 23% w/w is critical for its potential antiproliferative activity as the negative charges introduced to the EPS enable the interaction with the cell targets to trigger apoptosis. The activity is very specific to the leukemic T cell lineage, hence finds a prospective application in therapy with reduced side effects [62].

The process of emulsion polymerization is adopted to load Fe_3O_4 nanoparticles to the EPS produced by *Bifidobacterium*. This composite could induce apoptosis in the human gastric cancer cell lines transplanted into nude mice. TUNEL (terminal deoxynucleotidyl transferase dUTP nick end labeling) assay, immunohistochemistry and TEM (transmission electron microscope) of the transplant confirm the induction of apoptosis of the tumor cells in vivo [63].

TABLE 8.4   Chemical Composition of Some Bacterial EPS having Anticancer Activities

Bacteria	Chemical composition of the EPS
Lactobacillus acidophilus LMG9433[T]	Anionic heteropolymer of glucose, galactose, glucuronic acid, and 2-acetamido-2-deoxy-d-glucose [64].
Rhizobium N613	β-glucan type homopolysaccharide with linear 1, 4 linkage and branched 1,6 linkages [61].
Halomonas stenophila B100[T]	Anionic heteropolymer of glucose, galactose and mannose, functionalized with uronic acid, sulfate and phosphate groups, MW 3.75 ′ 10^5 Da [62].

8.2.4 *EXOPOLYSACCHARIDES AS ANTICOAGULANT AGENTS*

Sulfated linear polysaccharides like heparin sulfate play a crucial role in the blood coagulation cascade. Heparin sulfate is found in all animal tissues, however, when required to be used as a therapeutic, the extraction from animal tissues becomes a concern with respect to its pharmacokinetic compatibility and risk of contamination with prions. Several marine macroalgae are reported to produce sulfated EPS, which concurrently have anticoagulant activity [65]. Anticoagulant activity of a sulfated polysaccharide is attributed to multiple factors like degree and position of sulfate groups, chain length of the EPS and the type of glycosidic linkage. The EPS produced by some marine bacteria are also known to be sulfated; however, there are no reports till date for native bacterial EPS having anticoagulant activity. Hence, chemical sulfation is adopted to enhance the anticoagulant activity of some selected EPS.

Chemical modifications particularly sulfation and depolymerization impart anticoagulant activity to the EPS produced by *Alteromonas infernus*. The EPS is sulfated sulfur trioxide pyridine complex and depolymerised by free radical depolymerization to yield low molecular weight EPS fractions with about 40% w/w sulfate substitution. The increased sulfate ester content at the 4-O position lowers the thrombin generation via contact activation and thromboplastin activation systems and thus increases anticoagulant activity of the EPS [66]. Further research directed towards the identification of sulfated polysaccharides or chemical modification of some native EPS can lead to the development of newer and safer anticoagulant agents from bacterial origin.

8.2.5 EXOPOLYSACCHARIDES IN TISSUE ENGINEERING AND REGENERATIVE MEDICINE

The field of tissue engineering and regenerative medicine aims at promoting the regeneration of tissues or replacement of failing or malfunctioning organs, by means of combining a scaffold or support material, adequate cells and bioactive molecules. Biopolymers have emerged as forerunners in such applications due to their biocompatibility [67, 68] (Table 8.5).

TABLE 8.5 Chemical Composition of Some Bacterial EPS having Prospective Applications in Tissue Engineering and Regenerative Medicine

Bacteria	Chemical composition of the EPS
Vibrio diabolicus	Heteropolymer of glucuronic acid, N-acetylglucosamine and N-acetyl galactosamine, MW 8 ´ 10^5 Da [72].
Alteromonas infernus GY785^T	Heteropolymer of glucose, galactose, galacturonic acid, glucuronic acid and rhamnose, functionalized with sulfates, MW ~10^6 Da [16].
Alteromonas macleodii	Heteropolymer of glucose, galactose, mannose, galacturonic acid and pyruvated mannose, MW 3.3 ´ 10^5 Da [74].

The culture supernatant of *Lactobacillus plantarum* containing EPS is mixed with eucerin (preservative) and applied as a wound dressing on incised wound created in a mouse model. The probiotic treated group shows a marked acceleration in wound healing by initial macrophage infiltration and fibrobast proliferation; and decrease inflammation by the gradual lowering of the neutrophils at the wound site [69].

    Thin film of bacterial cellulose isolated from *Acetobacter xylinus* induces proliferation, spreading and migration of human keratinocytes in vitro and can be used as a therapeutic agent for healing skin wounds [27].

Gellan gum based microparticles reinforced with acryl solution, form hydrogels capable of proliferating the growth of nucleus pulposus (structure which undergoes initial degradation of the intervertebral disc) in vitro. Hence, it is proposed to be a candidate strategy for inter vertebral disc regeneration [31].

Recently, two EPS extracted from marine bacteria; HE800 (an acetylated EPS produced by *Vibrio diabolicus*) or GY785 (a sulfated EPS produced by *Alteromonas infernus*) were used to prepare injectable hydrogels with silylated hydroxypropylmethycellulose (sHPMC) [70]. These hydrogels showed proliferative activity in osteoblast cell line (MC3T3-1) and chondrocyte cell line (C28/l2) in two and three-dimensional matrices suggesting their use as scaffold for bone materials. This property is attributed to the negatively charged groups like sulfate and acetyl on the EPS which can react with the cationic proteins associated with growth factors, cytokines and cell adhesion molecules [70]. The EPS, HE800 EPS is functionally comparable to hyaluronic acid and shows strong bone healing properties in experimental rats [71]. It also shows proliferative activity on nonmineralised connective tissues like skin, gum, cartilage and tendon [72].

Chemical sulfation of EPS produced by *Alteromonas infernus* favors bone resorption by inhibiting osteoclastogenesis and accelerating differentiation of osteoblasts in vitro in bone marrow stem cell (BMSC) and murine RAW 264.7 cell lines. The activity is due to the sulfate groups incorporated in the EPS, which enable the formation of a heteromolecular complexation with necrotic factor (NF-κB). This sulfated EPS depolymerised by free radical depolymerization shows proliferative effects on human umbilical vein endothelial cells (HUVEC) in vitro. It also shows low anticoagulant activity and hence can be used to accelerate vascular wound healing [16]. It is patented as a wound healing material for the connective tissues, particularly skin and gum tissues [73].

8.2.6 EXOPOLYSACCHARIDES IN DRUG DELIVERY

Certain polysaccharides can form three-dimensional, hydrophilic, polymeric networks by chemical or physical cross-link to form structures called hydrogel. Hydrogels are capable of imbibing large amounts of water or biological fluids. This property together with the poor solubility of some EPS in biological fluids has enabled their usage in drug delivery systems.

Dextran is used extensively in colonic drug delivery systems. The abundant -OH groups in the polymer readily link to the drug molecules. Enzymes required for the digestion of dextran are not synthesized in the stomach and small intestine hence the drug remains intact till it reaches the colon. The colonic microflora produces enzyme dextranase, which cleave dextran and release the drug in the colon [75]. Xanthan produced by *Xanthomonas campestris,* is used for its application as an ophthalmic drug delivery system. Ophthalmic drug echothiophate iodide (used for treatment of glaucoma) when immobilized in xanthan can alleviate the side effects of the drug by controlled slow release and thereby enable the usage of a lower drug dose [39]. The pregranulated formulation of xanthan is patented for delayed drug release of pharmaceutical agents in the form of tablets [76]. Xanthan is also used for the preparation of sponge like in situ gelling inserts for the delivery of proteins and peptides in the nasal cavity [77].

Alginate has in situ gelling property due to the presence of L-guluronic acid. Ocular drug formulations containing alginate isolated from bacteria readily form gel in stimulated lachrymal fluid enabling prolonged release of an ocular drug pilocarpine [78]. Oral administration of a pharmaceutical formulation containing aqueous alginate, drug (paracetamol) calcium ions and sodium citrate forms gel in the acidic pH of the stomach and enables the slow release of drug thus reducing the possible side effects [79].

Mauran sulfate an EPS produced by *Halomonas maura* is an anionic heteropolysaccharide of glucose, mannose, galactose and glucuronic acid with functional sulfate and phosphate groups. Average MW of the polymer is 4.7×10^6 Da [80]. A stable nanocomposite biomaterial of anionic mauran sulfate and cationic chitosan is prepared by polyelectrolyte complexation. The composite enables sustained release of the drug, 5-flurouracil (drug against breast adenocarcinoma) up to 12 days. The composite finds a prospective application in pharmaceutics and biomedical technology [23]. Thus, the differential gelling properties of the EPS and the ease for drug conjugation can be further explored for the development of targeted and prolonged drug delivery systems.

8.2.7   COSMECEUTICAL APPLICATIONS OF EXOPOLYSACCHARIDES

The inherent properties of the EPS to maintain a hydrated matrix along with the therapeutic and emulsifying properties enable the application as a therapeutic, cosmetic stabilizer and emulsifier.

The EPS produced by a bacterium *Alteromonas macleodii* is patented in the name "Deepsane" and the *Alteromonas macleodii* ferment (processed cell free supernatant) is a major component of a patented formulation for skin care applications named Abyssine ® 657 by Atrium laboratories. This EPS can alleviate the expression of ICAM-1(Intercellular Adhesion Molecule) in keratinocytes and elevate keratinocyte proliferation hence is used to protect irritable, hyper-reactive or sensitive skin from inflammation [81]. The EPS is also used as an active ingredient in some other patented skin care formulations [82, 83].

An EPS isolated from the bacteria from hydrothermal origin [20] and nine different polysaccharides from bacteria isolated from Polynesian microbial mats [84] are patented for preparation of several cosmetic formulations. The EPS is selected based on the therapeutic response shown in vitro. Each formulation prepared, contains at least one EPS as an active component. The EPS has protective effects on UV-B induced damage to the Langerhans cells; it up regulates the secretion of IL-α by cultured keratinocytes and enhances the phagocytic activity of the keratinocytes against acne causing bacteria. β-1, 3 polyglucan like EPS produced by *Cellulomonas flavigena* is patented for use in cosmeceutics as it can control the supply and loss of water to and from stratum corneum layer of the skin. This property helps in enhancing or prolonging skin moisturization and also acts as a cosmetic stabilizer [85].

The cosmeceutical market is the fastest growing segment in the personal care industry. Research focused on identification of newer EPS from bacteria, exploration of its properties towards skin protection and validation of the activities in vivo can revolutionize the application of EPS in cosmetics.

8.2.8   FUNCTIONAL FOODS

The general category of functional foods includes processed food or foods fortified with health-promoting additives. In this respect the EPS produced by lactic acid bacteria (LAB) such as *Lactococcus lactis, Lactobacillus casei, Lactobacillus sake, Lactobacillus rhamnosus* and yogurt bacteria, which are *Streptococcus thermophilus* and *Lactobacillus delbrueckii* ssp. *bulgaricus* etc., are extensively being used in fermented dairy products. In particular, for the production of yogurt, drinking yogurt, cheese, fermented cream, milk-based desserts etc. The food industry uses polysaccharides as thickeners, emulsifiers, gelling agents and stabilizers. Although EPS does not have any flavor, EPS produced by some *Lactobacillus* spp. of bacteria enhances the mouth feel of the fermented products by improved volatilization of the intrinsic flavors [86].

In addition to flavor benefits certain EPS produced by LAB are also claimed to have beneficial effects on health as anticancer and immune-modulatory agents as evidenced by in vivo studies [87]. Hence, the future will show a shift from pure technical and texturing applications of selected EPS to more targeted improved consumer health benefits [88].

8.3   PRODUCTION OF EPS

The structural complexity of EPS is genetically controlled and is unique to the bacterial species producing it. Hence, considering the taxonomic diversity of bacterial species, the structural diversity of EPS produced by them is also large and expanding. Environmental habitat of the bacteria often influences the nature of the EPS produced e.g., bacteria isolated from hypersaline environments possess functional groups such as sulfate as an adaptation to osmotic stress. Accordingly, when grown under laboratory conditions the chemistry of the EPS produced by bacteria can be tailored to some extent by physiological control by regulating growth temperature and pH. Exopolysaccharide production rate and the chemical composition of the EPS depends on several critical factors like the media composition in terms of carbon and nitrogen source, mineral requirements, availability of micronutrients etc. Increased EPS production is observed in some marine bacteria when grown on limited nutrients (such as nitrogen, phosphorus, sulfur and potassium). Suboptimal temperature for growth, osmotic stress, and presence of antibiotics, dehydrating agents like ethanol or other physical factors that restrict the bacterial growth may enhance the EPS production [89]. The choice of selected carbon source in the growth medium represents the first step in the optimization of EPS production. Enhanced EPS production is attained in some selected Pseudomonads grown in sucrose containing media when subjected to physiological stress like dehydration by ethanol and increased osmolarity [90]. Overproduction studies on EPS by *Rhizobium meliloti* strains have shown that phosphate limitation [91] and increased manganese [92] are some factors that induce EPS production. An EPS overproducing mutant is obtained after exposure of *Rhizobium trifolii* sp. to UV irradiation [93]. Emulsan production is stimulated by *Acinetobacter calcoaceticus* when grown in the presence of protein synthesis inhibitory antibiotics like chloramphenicol or amino acid deprivation [94].

Bacteria often produce more than one EPS that may be having the same chemical composition but varying molecular weights as in *Bacillus thermoantarcticus* [95] or sometimes may produce two EPS with varied chemical composition along with varying molecular weight and different gelling properties as in *Alteromonas* sp. [96].

Development of efficient fermentation technologies will uplift the usage of many more potential bacterial EPS at an industrial scale and open up a new avenue for polysaccharide research.

8.4 CONCLUSIONS

Several applications of bacterial EPS have been discussed; however, research is in its infancy due to the need for multidisciplinary knowledge and expertise to translate the biological activities of EPS to therapeutic applications. Quite adversely, for applications like tissue scaffolding and drug delivery most of the studies use structurally well-characterized and commercially available EPS like dextran, xanthan, alginate etc., cornering the search for novel polysaccharides. Newly elucidated EPS can be commercialized only if; basic research on structural identification of EPS goes parallel with the advanced applied research.

The avenues for bacterial EPS in therapeutics and cosmetics are highly promising as the structural specification and related pharmaceutical activity of the bacterial EPS surpasses the production cost and yield issues. Advances in the understanding of the biosynthesis, structure and related function of EPS from bacterial origin have shown the crucial role of these molecules in modulating disease and maintaining health. There is a growing need for developing a holistic strategic approach in developing and bioprospecting EPS as biocompatible and efficient therapeutics.

KEYWORDS

- **Exopolysaccharides**
- **Bacterial**
- **Bioprospecting**
- **Carbohydrate polymers**
- **Bacterial exopolysaccharides**
- **Secondary metabolites**
- **Dextran**
- **Alginate**

REFERENCES

1. Nwodo, U. U., Green, E., Okoh, A. I. (2012). Bacterial exopolysaccharides: Functionality and prospects. Int. J. Mol Sci., 13, 14002–14015.

2. Tsuneda, S., Aikawa, H., Hayashi, H., Yuasa, A., & Hirata, A. (2003). Extracellular Polymeric Substances Responsible for Bacterial Adhesion onto solid surface, FEMS Microbiol Lett. 223(2), 287–292.

3. Rehm, B. (2009). Microbial Production of Biopolymers and Polymer Precursors, applications and perspectives, Caister Academic Press Norwich.

4. Kenne, L., & Lindberg, B. (1983). Bacterial polysaccharides. In The Polysaccharides, Aspinall, G. O. Ed., Academic Press New York, 2, 287–363.

5. Ito, T., Higuchi, T., Hirobe, M., Hiramatsu, K., & Yokota, T. (1994). Identification of a novel sugar, 4-amino-4, 6-dideoxy-2-O-methylmannose in the lipopolysaccharide of Vibrio cholerae O1 Serotype Ogawa, Carbohyd Res 256(1), 113–128.

6. Sharon, N. (2007). Celebrating the Golden Anniversary of the Discovery of Bacillosamine the Diamino Sugar of a Bacillus Glycobiology, 17(11), 1150–1155.

7. Sutherland, I. W. (2001). Biofilm exopolysaccharides a strong and sticky framework, Microbiology, 147, 3–9.

8. Ridout, M. J., Brownsey, G. J., York, G. M., Walker, G. C., & Morris, V. J. (1997). Effect of O-acyl substituents on the functional behavior of Rhizobium meliloti succinoglycan Int. J. Biol Macromol, 20(1), 1–7.

9. Hassler, R. A., & Doherty, D. H. (2008). Genetic Engineering of Polysaccharide structure: production of variants of xanthan gum in Xanthomonas campestris Biotechnol Progr. 6(3), 182–187.

10. Smidsrod, O., & Haug, A. (1968). Dependence upon uronic acid composition of some ion-exchange properties of alginates, Acta Chem. Scand, 22(6), 1989–1997.

11. Velasco, C. R., Baud'Huin, M., Sinquin, C., Maillasson, M., Heymann, D., Colliec-Jouault, S., & Padrines, M. (2011). Effects of a sulfated exopolysaccharide produced by Altermonas infernus on bone biology. Glycobiology, 21(6), 781–795.

12. Kitazawa, H., Harata, T., Uemura, J., Saito, T., Kaneko, T., & Itoh, T. (1998). Phosphate group requirement for mitogenic activation of lymphocytes by an extracellular phosphopolysaccharide from Lactobacillus delbrueckii ssp bulgaricus. Int. J. Food Microbiol, 40(3), 169–175.

13. MacCormick, C. A., Harris, J. E., Jay, A. J., Ridout, M. J., Colquhoun, I. J., & Morris, V. J. (1996). Isolation and characterization of a new extracellular polysaccharide from an Acetobacter species. J. Appl. Microbiol 81(4), 419–424.

14. Rehm, B. H. A. (2010). Bacterial polymers, Biosynthesis, modifications and applications. Nat. Rev.Microbiol, 8, 578–592.

15. Xu, C. L., Wang, Y. Z., Jin, M. L., & Yang, X. Q. (2009). Preparation characterization and immunomodulatory activity of selenium-enriched exopolysaccharide produced by bacterium Enterobacter cloacae Z0206, Bioresource Technol, 100, 2095–2097.

16. Matou, S., Colliec, J. S., Galy, F. I., Ratiskol, J., Sinquin, C., Guezennec, J., Fisher, A.M., & Helley, D. (2005). Effect of an oversulfated exopolysaccharide on angiogenesis induced by fibroblast growth factor-2 or vascular endothelial growth factor in vitro. Biochem. Pharmacol, 69(5), 751–759.

17. Velasco, C. R., Baud'Huin, M., Duplomb, L., Colliec-Jouault, S., Heymann, D., & Padrines, M. (2009). Effects of a Sulphated exopolysaccharide produced by Alteromonas infernus on rat bone cells Bone, 44(2), S296.

18. Arena, A., Gugliandolo, C., Stassi, G., Pavone, B., Iannello, D., Bisignano, G., & Luciana, T. M. (2009). An exopolysaccharide produced by Geobacillus thermodenitrificans strain B3-72: Antiviral activity on immunocompetent cells, Immunol. Lett. 123(2), 132–137.

19. Dugan, J. M., Gough, J. E., & Eichhorn S. J. (2013). Bacterial cellulose scaffolds and cellulose nanowhiskers for tissue engineering. Nanomedicine, 8(2), 287–298.

20. Vacher, A. M., & Fritsch, M. C. (September 24, 2009). Cosmetic composition which includes at least one polysaccharide derived from bacteria of hydrothermal origin. United States Patent US2009/0238782 A1.

21. Sihorkar, V., & Vyas, S. P. (2001). Potential of Polysaccharide Anchored Liposomes in Drug Delivery, targeting and immunization. J. Pharm. Pharmaco. Sci., 4(2), 138–158.

22. Surekha, K. S., Ibrahim, M. B., Prashant, K. D., Arun, G. B., & Balu, A. C. (2010). Biosurfactants, bioemulsifiers and exopolysaccharides from marine microorganisms, Biotechnol Adv., 28(4), 436–450.

23. Raveendran, S., Poulose, A. C., Yoshida, Y., Maekawa, T., & Kumar, D. S. (2013). Bacterial exopolysaccharide based nanoparticles for sustained drug delivery, cancer chemotherapy and bioimaging, Carbohyd. Polym, 91(1), 22–32.

24. Iyer, A., Mody, K., & Jha, B. (2004). Accumulation of hexavalent chromium by an exopolysaccharides producing marine Enterobacter cloacae. Mar. Pollut. Bull., 49, 974–977.

25. Das, P., Mukherjee, S., & Sen, R. (2009). Biosurfactant of marine origin exhibiting heavy metal remediation properties. Bioresource Technol, 100, 4887–4890.

26. Touzel, J. P., Chabbert, B., Monties, B., Debeire, P., & Cathala, B. (2003). Synthesis and characterization of dehydrogenation polymers in Gluconacetobacter xylinus cellulose and cellulose/pectin composite. J. Agr. Food Chem., 51(4), 981–986.

27. Sanchavanakit, N., Sangrungraungroj, W., Kaomongkolgit, R., Banaprasert, T., Pavasant, P., & Phisalaphong, M. (2006). Growth of human keratinocytes and fibroblasts on bacterial cellulose film. Biotechnol Progr., 22(4), 1194–1199.

28. Vu. B., Chen, M., Crawford, R. J., & Ivanova, E. P. (2009). Bacterial Extracellular polysaccharides involved in biofilm formation, Molecules, 14(7), 2535–2554.

29. Mehvar, R. (2000). Dextrans for targeted and sustained delivery of therapeutic and imaging agents. J. Control. Release, 69(1), 1–25.

30. Harding, N., Patel, Y., & Coleman, R. (2004). Organization of genes required for gellan polysaccharide biosynthesis in Sphingomonas elodea ATCC 31461. J. Ind. Microbiol Biot, 31(2), 70–82.

31. Pereira, D. R., Silva-Correia, J., Caridade, S. G., Oliveira, J. T., Sousa, R. A., Salgado, A. J., Joaquim M. O., João F. M., Sousa, N., & Reis, R. L. (2011). Development of gellan gum-based microparticles/hydrogel matrices for application in the intervertebral disc regeneration. Tissue Engineering Part C: Methods, 17(10), 961–972.

32. Harada, T., Misaki, A., & Saito, H. (1968). Curdlan A bacterial gel-forming β-1, 3-glucan. Arch. Biochem. Biophys, 124, 292–298.

33. Provonchee, R. B., & Renn, D. W. (September 27, 1988). Polysaccharide compositions, preparation and uses. U.S. Patent 4, 774,093.

34. Sabra, W., Zeng, A. P., & Deckwer, W. D. (2001). Bacterial Alginate Physiology, product quality and process aspects. Appl. Microbial. Biot. 56(3), 315–325.

35. Knill, C. J., Kennedy, J. F., Mistry, J., Miraftab, M., Smart, G., Groocock, M. R., & Williams, H. J. (2004). Alginate fibers modified with unhydrolyzed and hydrolyzed chitosans for wound dressings. Carbohyd. Polym, 55(1), 65–76.

36. Jansson, P. E., Lindberg, B., Widmalm, G., & Sandford, P. A. (1985). Structural studies of an extracellular polysaccharide (S-130) elaborated by Alcaligenes ATCC 31555. Carbohydr. Res., 139, 217–223.

37. Ribier, A., & Simonnet, J. T. (April 18, 2000). Process for the stabilization of vesicles of amphiphilic lipid (s) and composition for topical application containing the said stabilized vesicles. U. S. Patent 6,051,250.

38. Jansson, P. E., Kenne, L., & Lindberg, B. (1975). Structure of the extracellular polysaccharide from Xanthomonas campestris. Carbohydr Res., 45(1), 275–282.

39. Lin, S. L., & Pramoda, M. K. (January 23, 1979). Xanthan gum therapeutic compositions. U. S. Patent 4,136,177.

40. Yoshida, S., Nakamura, T., Ishikubo, A., & Nasu, A. (July 29, 2004). Cosmetic Oil-in-Water emulsion preparation, WIPO Patent WO/2004/062631.

41. Halliwell, B., Gutteridge, J., & Aruoma, O. I. (1987). The deoxyribose method a simple "test-tube" assay for determination of rate constants for reactions of hydroxyl radicals. Anal. Biochem, 165(1), 215–219.

42. Balasubramanian, B., Pogozelski, W. K., & Tullius, T. D. (1998). DNA strand breaking by the hydroxyl radical is governed by the accessible surface areas of the hydrogen atoms of the DNA backbone. P. Natl. A. Sci., 95(17), 9738–9743.

43. Zhang, L., Liu, C., Li, D., Zhao, Y., Zhang, X., Zeng, X., Yang Z., & Li, S. (2013). Antioxidant activity of an exopolysaccharide isolated from Lactobacillus plantarum C88. Int. J. Biol. Macromol, 54, 270–275.

44. Kodali, V. P., & Sen, R. (2008). Antioxidant and free radical scavenging activities of an exopolysaccharide from a probiotic bacterium. Biotechnol J., 3(2), 245–251.

45. Kanmani, P., Yuvaraj, N., Paari, K. A., Pattukumar, V., & Arul, V. (2011). Production and purification of a novel exopolysaccharide from lactic acid bacterium Streptococcus phocae PI80 and its functional characteristics activity in vitro. Bioresource Technol, 102(7), 4827–4833.

46. Guo, S., Mao, W., Han, Y., Zhang, X., Yang, C., Chen, Y., Chen, Y., Xu, J., Li, H., Qi, X., & Xu, J. (2010). Structural characteristics and antioxidant activities of the extracellular polysaccharides produced by marine bacterium Edwardsiella tarda Bioresource Technol, 101, 4729–4732.

47. Tallon, R., Bressollier, P., & Urdaci, M. C. (2003). Isolation and characterization of two exopolysaccharides produced by Lactobacillus plantarum EP56. Res. Microbiol, 154(10), 705–712.

48. Kodali, V. P., Das, S., & Sen, R. (2009). An exopolysaccharide from a probiotic: Biosynthesis dynamics, composition and emulsifying activity. Food Res. Int., 42(5), 695–699.

49. Moskovitz, J., Yim, M. B., & Chock, P. B. (2002). Free radicals and disease. Arch. Biochem. Biophys, 397(2), 354–359.

50. Im, S. A., Wang, W., Lee, C. K., & Lee, Y. N. (2010). Activation of macrophages by exopolysaccharide produced by MK1 bacterial strain isolated from neungee mushroom, Sarcodon aspratus. Immune Netw. 10(6), 230–238.

51. Kitazawa, H., Yamaguchi, T., Miura, M., Saito, T., & Itoh, T. (1993). B-Cell Mitogen produced by slime-forming, encapsulated Lactococcus lactis ssp. cremoris isolated from ropy sour milk, Viili. J. Dairy Sci., 76(6), 1514–1519.

52. Suzuki, C., Kimoto-Nira, H., Kobayashi, M., Nomura, M., Sasaki, K., & Mizumachi, K. (2008). Immunomodulatory and cytotoxic effects of various Lactococcus strains on the murine macrophage cell line J774.1. Int. J Food Microbiol, 123(1), 159–165.

53. Chabot, S., Yu, H. L., De Léséleuc, L., Cloutier, D., Van Calsteren, M. R., Lessard, M., Roy D., Lacrouix, M., & Oth, D. (2001). Exopolysaccharides from Lactobacillus rhamnosus RW-9595 M stimulate TNF, IL-6 and IL-12 in human and mouse cultured immunocompetent cells, and IFN-in mouse splenocytes. Le Lait. 81(6), 683–697.

54. Arena, A., Maugeri, T. L., Pavone, B., Iannello, D., Gugliandolo, C., & Bisignano, G. (2006). Antiviral and immunoregulatory effect of a novel exopolysaccharide from a marine thermotolerant Bacillus licheniformis. Int. Immunopharmacol, 6(1), 8–13.

55. Makino, S., Ikegami, S., Kano, H., Sashihara, T., Sugano, H., Horiuchi, H., Saito, T., & Oda, M. (2006). Immunomodulatory Effects of polysaccharides produced by Lactobacillus delbrueckii ssp. bulgaricus OLL1073R-1. J Diary Sci. 89(8), 2873–2881.

56. Macedo, M. G., Lacroix, C., Gardner, N. J., & Champagne, C. P. (2002). Effect of medium supplementation on exopolysaccharide production by Lactobacillus rhamnosus RW-9595 M in whey permeate. Int. Dairy J. 12(5), 419–426.

57. Nicolaus, B., Panico, A., Manca, M. C., Lama, L., Gambacorta, A., Maugeri, T., Gugliandolo, C., & Caccamo, D. A. (2000). Thermophilic Bacillus isolated from an Eolian shallow hydrothermal vent able to produce exopolysaccharides. Syst. Appl. Microbiol., 23(3), 426–432.

58. Zong, A., Cao, H., & Wang, F. (2012). Anticancer polysaccharides from natural resources: A review of recent research. Carbohyd. Polym, 90(4), 1395–1410.

59. Kim, Y., Oh, S., Yun, H. S., & Kim, S. H. (2010). Cell-bound exopolysaccharide from probiotic bacteria induces autophagic cell death of tumor cells. Lett. Appl. Microbiol, 51(2), 123–130.

60. Chen, X., Jiang, H., Yang, Y., & Liu, N. (2009). Effect of exopolysaccharide from Bifidobacterium bifidum on cell of gastric cancer and human telomerase reverse transcriptase. Acta Microbiol. Sin, 49(1), 117–122.

61. Zhao, L., Chen, Y., Ren, S., Han, Y., & Cheng, H. (2010). Studies on the chemical structure and antitumor activity of an exopolysaccharide from Rhizobium sp. N613. Carbohyd Res., 345(5), 637–643.

62. Ruiz, R. C., Srivastava, G. K., Carranza, D., Mata, J. A., Llamas, I., Santamaria, M., Quesada, E., & Molina, I. J. (2011). An exopolysaccharide produced

by the novel halophilic bacterium Halomonas stenophila strain B100 selectively induces apoptosis in human T leukemia cells. Appl. Microbiol Biot. 89, 345–355.

63. Zhang, Y., Li, Y. T., Hou, R. Q., & Liu, N. (2008). Bifidobacterium exopolysaccharide loaded nanoparticles induces apoptosis of human gastric cancer cells transplanted into nude mice, Chinese J. Cancer Biotherapy, 2, 1–9.

64. Robijn, G. W., Gutiérrez Gallego, R., van den Berg, D. J., Haas, H., Kamerling, J. P., & Vliegenthart, J. F. (1996). Structural characterization of the exopolysaccharide produced by Lactobacillus acidophilus LMG9433 Carbohyd Res., 288, 203–218.

65. Majdoub, H., Ben Mansour, M., Chaubet, F., Roudesli, M. S., & Maaroufi, R. M. Anticoagulant activity of a sulfated polysaccharide from the green alga Arthrospira platensis. Biochim. Biophys. Acta, 1790, 1377–1381.

66. Colliec -Jouault, S., Chevolot, L., Helley, D., Ratiskol, J., Bros, A., Sinquin, C., Roger O., & Fischer, A. M. (2001). Characterization, chemical modifications and in vitro anticoagulant properties of an exopolysaccharide produced by Alteromonas infernus. BBA. Gen. Subjects 1528(2), 141–151.

67. Mano, J. F., Silva, G. A., Azevedo, H. S., Malafaya, P. B., Sousa, R. A., Silva, S. S., Boesel, L. F., Oliveira, J. M., Santos, T. C., Marques, A. P., Neves, N. M., & Reis, R. L. (2007). Natural origin biodegradable systems in tissue engineering and regenerative medicine present status and some moving trends. J. R. Soc. Interface. 4(17), 999–1030.

68. Ko, H. F., Sfeir, C., & Kumta, P. N. (2010). Novel synthesis strategies for natural polymer and composite biomaterials as potential scaffolds for tissue engineering. Phil. Trans. R. Soc. A. 368(1917), 1981–1997.

69. Nasrabadi, H. M., Ebrahimi, T. M., & Banadaki D. S. (2011). Study of cutaneous wound healing in rats treated with Lactobacillus plantarum on days 1, 3, 7, 14 and 21. Afr. J. Pharm. Pharmacolo, 5(21), 2395–2401.

70. Rederstorff, E., Weiss, P., Sourice, S., Pilet, P., Xie, F., Sinquin, C., Colliec-Jouault S., Guicheux, J., Laïb, S. (2011). An in vitro study of two GAG-like marine polysaccharides incorporated into injectable hydrogels for bone and cartilage tissue engineering. Acta Biomater, 7(5), 2119–2130.

71. Zanchetta, P., Lagarde, N., & Guezennec, J. (2003). A new bone-healing material a hyaluronic acid-like bacterial exopolysaccharide, Calcified Tissue Int., 72(1), 74–79.

72. Guezennec, J., Pignet, P., Raguenes, G., & Rougeaux, H. (August 20, 2002). Marine bacterial strain of the genus vibrio, water-soluble polysaccharides produced by said strain and their uses. U. S. Patent No. 6,436,680.

73. Senni, K., Pereira, J., Gueniche, F., Delbarre-Ladrat, C., Sinquin, C., Ratiskol, J., Godeau, G., Fischer, A. M., Helley, D., & Colliec-Jouault, S. (2011). Marine polysaccharides: a source of bioactive molecules for cell therapy and tissue engineering. Mar Drugs, 9(9), 1664–1681.

74. Raguenes, G., Pignet, P., Gauthier, G., Peres, A., Christen, R., Rougeaux, H., Barbier, G., & Guezennec, J. (1996). Description of a new polymer-secreting bacterium from a deep-sea hydrothermal vent, Alteromonas macleodii subsp.

fijiensis, and preliminary characterization of the polymer. Appl. Environ. Microb, 62(1), 67–73.

75. Drasar, B. S., Hill, M. J. (1974). The distribution of bacterial flora in the intestine, In: Human Intestinal Flora. Academic Press, New York, 15–43.

76. Collaueri, J. P., Conrath, G., Derian, P. J., Gousset, G., & Mauger, F. (April 24, 2001). Pharmaceutical compositions in the form of sustained-release tablets based on high molecular weight polysaccharide granules. U. S. Patent, 6,221,393.

77. Bertram, U., & Bodmeier, R. (2006). In situ gelling, bioadhesive nasal inserts for extended drug delivery, in vitro characterization of a new nasal dosage form Eur. J. Pharm. Sci., 27, 62–71.

78. Cohen, S., Lobel, E., Trevgoda, A., & Peled, Y. (1997). A novel in situ-forming ophthalmic drug delivery system from alginates undergoing gelation in the eye, J. Control Release, 44, 201–208.

79. Kubo, W., Miyazaki, S., & Attwood, D. (2003). Oral sustained delivery of Paracetamol from in situ-gelling gellan and sodium alginate formulations, Int. J. Pharm, 258, 55–56.

80. Arias, S., Del Moral, A., Ferrer, M. R., Tallon, R., Quesada, E., Béjar, V., & Mauran, (2003). An exopolysaccharide produced by the halophilic bacterium Halomonas maura, with a novel composition and interesting properties for biotechnology, Extremophiles, 7, 319–326.

81. Thibodeau, A. (2005). Protecting the skin from environmental stresses with an exopolysaccharide formulation Cosmet Toiletries., 120(12), 81–86.

82. Kim, H. J., Jung, S. W., Kim, H., Kim, S. J., Lee, J. D., Ryoo, H. C., Kwon A.Y., & Jo, M. R. (May 27, 2004). Whitening and antionxidative cosmetic composition containing resveratrol and method for preparing the same. U. S. Patent Application 10/558,348.

83. Dupont, E., Samson, M., & Galderisi, A. (October 14, 2010). Skin care compositions and method of use thereof. U. S. Patent Application 12/904,263.

84. Loing, E., Briatte, S., Vayssier, C., Beaulieu, M., Dionne, P., Richert, L., & Moppert, X. (December 29, 2010). Cosmetic compositions comprizing exopolysaccharides derived from microbial mats, and use thereof. European Patent EP 2265249.

85. Davis, W. B. (October 27, 1992). Unique bacterial polysaccharide polymer gel in cosmetics, pharmaceuticals and foods. U.S. Patent No. 5,158,772.

86. Duboc, P., & Mollet, B. (2001). Applications of exopolysaccharides in the dairy industry. Int. Dairy J., 11(9), 759–768.

87. Patel, S., Majumder, A., & Goyal, A. (2012). Potentials of exopolysaccharides from lactic acid bacteria. Indian J. Microbiol., 52(1), 3–12.

88. Vincent, S. J., Faber, E. J., Neeser, J. R., Stingele, F., & Kamerling, J. P. (2001). Structure and properties of the exopolysaccharide produced by Streptococcus macedonicus Sc136 Glycobiology, 11, 131–139.

89. Sutherland, I. W. (1982). Biosynthesis of microbial exopolysaccharides, Adv. Microbial Physiol., 23, 79–150.

90. Singh, S., & Fett, W. F. (2005). Stimulation of Exopolysaccharide production by fluorescent pseudomonads in sucrose medium due to dehydration and increased osmolarity, FEMS Microbiol, Lett., 130(2), 301–306.

91. Zhan, H. J., Lee, C. C., & Leigh, J. A. (1991). Induction of the second exopolysaccharide (EPSb) in Rhizobium meliloti SU47 by low phosphate concentrations, J. Bacteriol, 173(22), 7391–7394.

92. Appanna, V. D. (1988). Stimulation of exopolysaccharide production in Rhizobium meliloti JJ-1 by manganese. Biotechnol Lett., 10(3), 205–206.

93. Ghai, J., Ghai, S. K., & Kalra, M. S. (1985). Ultraviolet-irradiation induced and spontaneous mutation of Rhizobium trifolii 11B in relation to water-soluble and water-insoluble polysaccharide production ability, Enzyme Microb Tech, 7(2), 83–86.

94. Rubinovitz, C., Gutnick, D. L., & Rosenberg, E. (1982). Emulsan production by Acinetobacter calcoaceticus in the presence of chloramplenicol, J. Bacteriol, 152(1), 126–132.

95. Manca, M. C., Lama, L., Improta, R., Esposito, E., Gambacorta, A., & Nicolaus, B. (1996). Chemical composition of two exopolysaccharides from Bacillus thermoantarcticus Appl. Environ. Microb, 62(9), 3265–3269.

96. Samain, E., Milas, M., Bozzi, L., Dubreucq, M., & Rinaudo, M. (1997). Simultaneous Production of two Different Gel-forming Exopolysaccharides by an Alteromonas strain originating from deep-sea hydrothermal vents. Carbohyd Polym., 34, 235–241.

CHAPTER 9

PREPARATION AND PROPERTIES OF COMPOSITE FILMS FROM MODIFIED CELLULOSE FIBER-REINFORCED WITH DIFFERENT POLYMERS

S. SANDEEP, LAXMISHWAR and G. K. NAGARAJA

CONTENTS

Abstract 170

9.1   Introduction... 170

9.2   Materials and Methodology 172

9.3   Results and Discussion 177

9.4   Conclusions... 191

Keywords 192

References.. 192

ABSTRACT

Ecofriendly concerns have led to an increased demand for good biodegradable polymer material with higher tensile strength and good gas barrier properties in many industrial applications and other material used in everyday life. Currently, almost all the polymers used do not physically decompose due to their high chemical stability, leading to the serious environmental problem. Now a day's replacing the nonbiodegradable polymeric material is an emerging interest field of research. Cellulose plays a very important role in this modification. Their low-cost and low density associated with high specific mechanical properties which is having biodegradable property alternative to the most commonly used synthetic reinforcement.

In this work, cellulose is modified by using 2-(Trifluoromethyl) benzoylchloride and 2-fluoro benzoylchloride by base catalyzed reaction and it is confirmed by its solubility and IR studies. The biodegradable composite films were developed by film casting method using modified cellulose with PLA or Poly (vinyl alcohol) or Polypyrrolidone in different compositions. The hybrid biodegradable composite films were developed by film casting method using modified cellulose with Poly (vinyl alcohol)-Poly (lactic acid); Poly (vinyl alcohol)-Polypyrrolidone in different compositions.

The film composites were characterized by mechanical, moisture absorption, gas barrier, and biodegradable properties. Obtained films have shown transparency and flexibility and displayed good mechanical properties. Film composites also showed good biodegradability.

9.1 INTRODUCTION

There is a growing worldwide interest pushed by governments and societies to increment the responsible use of renewable resources into commodity plastic products in order to reduce the waste associated with their use, particularly in packaging applications [1]. The use of biodegradable plastics and resources are seen as one of the many strategies to minimize the environmental impact of petroleum-based plastics. The biological base of these new biopolymers provides a unique opportunity to incorporate a highly demanded property of these materials, that is, the compostability. It must be considered that among the plastic waste there are products with a high degree of contamination and recycling requires a high-energy cost. Therefore, compostability is a very interesting property that guarantees that these new biomaterials will degrade mostly into carbon dioxide and water after disposal. These biodegradable materials present a number of excellent and promising properties in a number of applications, including packaging, automotive and biomedical sectors. Thus, thermoplastic biodegradable polymers, such as poly (lactic acid) (PLA), polyhydroxyalkanoate (PHA) and polycaprolactones (PCL), exhibit an excellent equilibrium of properties, that is, they are processable using conventional plastics machinery and, for the case of the first two, they arise from renewable resources. PLA is a thermoplastic biopolyester produced from L-lactid acid, which typically comes from the fermentation of corn starch. Currently, PLA is being commercialized and being used as a food packaging polymer for short shelf-life products with common applications such as containers, drinking cups, sundae, and salad cups, overwrap and lamination films, and blister packages.

In order to tailor the properties and reduce material costs, it is often desirable to combine bioplastic materials with other, ideally, more inexpensive substances, such as natural fibers. Reinforcement of some of these bioplastics with lignocellulosic fibers has previously been carried out with the overall aim of increasing its biodegradation rate and to enhance mechanical properties, that is, this route led to considerable improvements in the composites tensile strength [2].

The chemical modification of natural or synthetic polymers is an elaborate experimental task resulting in improvements in the original surface properties of the materials for use in academic and technological applications. From the viewpoint of the chemical potential of some natural or synthetic materials for useful applications, amorphous silica gel, chitosan, fruit peel, crystalline lamellar talc-like, clay-like or inorganic phosphate, and others have been explored. However, the richest natural raw material, cellulose, has been less exploited. This is despite the fact that this polysaccharide is encountered everywhere in nature, displays active potential functional groups available to react under appropriate conditions, displays changes in the surface after convenient immobilization, and can embody a diversity of functionalities. For modification, hydroxyl groups are available in the main skeleton and also in the branched chain, which is the most reactive position in the functionalization process. The applicability of such materials depends upon the availability of the active functions attached to the pendant chains covalently bonded to a given framework. As the original structure is changed into a surface, the hydrophilic character of the new material transforms its properties, leading eventually to a more hydrophobic behavior. Thus, these kinds of modified surface agents comprise a variety of organic molecules displaying amines, ketones, carboxylic acids, or thiol functions and also inorganic groups such as phosphates or oxides. Moreover, in general more than one functional basic center attached to the pendant molecule covalently bonded to the framework structure is desired.

However, cellulose has not reached its potential application in many areas because of its infusibility and insolubility. But at the same time, cellulosic fibers are hygroscopic in nature; moisture absorption can result in swelling of the fibers which may lead to microcracking of the composite and degradation of mechanical properties. This problem can be overcome by treating these fibers with suitable chemicals to decrease the hydroxyl groups, which may be involved in the hydrogen bonding within the cellulose molecules. Chemical treatments may activate these groups or can introduce new moieties that can effectively interlock with the matrix. A number of fiber surface treatments like silane treatment, benzylation and peroxide treatment were carried out which may result in improved mechanical performance of the fiber and composite. By limiting the substitution reaction on the surface of the fibers, good mechanical properties were obtained and a degree of biodegradability was maintained. As a result, various cellulose based composites have been prepared.

However, there is no literature regarding the combination of modified cellulose and polymers like PLA, polyvinyl alcohol etc., which is excellent mechanical properties and processing capabilities. To this end, biocomposites based on modified cellulose/different polymers were developed by a solution casting method in this work.

9.2 MATERIALS AND METHODOLOGY

9.2.1 MATERIALS

The fiber used in this work was commercial microcrystalline cellulose supplied by Loba Chemie. 2-(Trifluromethyl) benzoylchloride and Pyridine were purchased from Aldrich and used as received. The biopolymer of PLA (Mw is between 195,000 and 205,000 g/mol), Pyrrolidone, Poly (vinyl alcohol) used in this work was obtained from Cargill Dow LLC, Winibest Marketing Ltd, China and from FlukaAG (Switzerland), respectively. All the purchased chemicals used as such. The composite films were developed using Magic mold releasing agent and with Teflon mold of one square feet with 3 mm depth. The solvent acetonitrile was purchased from Rankem and used with purification. Finally the composite films were air dried in the hot air oven.

9.2.2 MODIFICATION OF CELLULOSE

Cellulose was treated with the saturated sodium hydroxide solution at room temperature and stirred for 2 hrs. After 2 hrs, solid obtained was filtered off. Salt formation was confirmed by solubility test, since it is freely soluble in water.

This salt was treated with 2-(Trifluromethyl) benzoylchloride in presence of pyridine as a base cum solvent and stirred overnight at 100 °C. Then dumped in to water; solid was filtered off. This product was confirmed by IR analysis, which shows the absence of peak at 3332 cm^{-1}. Figure 9.1 represents the IR spectra of cellulose and modified cellulose.

Cellulose Fiber OH + NaOH Cellulose Fiber O⁻ Na⁺ + H₂O

Cellulose Fiber O⁻ Na⁺ + R-COCl Cellulose Fiber OCO-R + NaCl

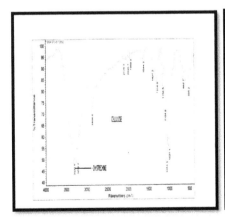
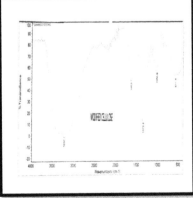

FIGURE 9.1   IR spectras of simple cellulose and modified cellulose.

9.2.3   PREPARATION OF FILMS

Modified cellulose was taken in a water with matrix phase composite (Poly(lactic acid), Poly (vinyl alcohol) or (Polypyrrolidone) in different composition like 10:90, 20:80, 30:70, 40:60, 50:50, 60:40, 70:30, 80:20, 90:10 and 95:05 ratio [3–6]. The reaction mixture was heated to 100 °C for 24 hrs. After 24 hrs. the reaction mass was turned to viscous state; it was allowed to room temperature and spread on the Teflon mold which was sprayed before by mold releasing spray and dried under vacuum oven at 100 °C to remove water contents completely. After complete drying, the films are stored in moisture free environment. The thickness of the films was in the range of 100µm to 350µm using Mitutoyo-7327 Dial Thickness Gage of accuracy 0.001 mm. Figure 9.2 represents the photograph of developed film.

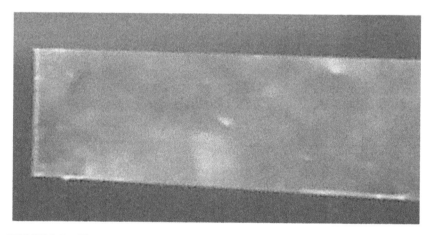

FIGURE 9.2   Photograph of developed film.

9.2.4   MOISTURE ABSORPTION EXPERIMENTS

From the composite sheets, all the specimens for moisture absorption experiments were cut with dimensions of 30 mm by 10 mm. Moisture absorption measurements were performed at 25 °C. Specimens were thoroughly washed and then vacuum dried until a constant weight was attained prior to the absorption experiments. At predetermined intervals, specimens were taken out from the chambers and weighed using a PGB200 model analytical balance.

The moisture uptake at any time points as a result of moisture absorption was determined by

$$\text{Moisture uptake} = \frac{Wh - Wo}{Wo} \times 100$$

where Wh and Wo denote the weight of humid specimens and the original dry value, respectively. All data from three repeated tests were averaged.

9.2.5 SOIL BURIAL DEGRADATION EXPERIMENTS

Under moisture controlled conditions soil burial degradation experiments were carried out at ambient temperature. Specimens of each composite were placed in a series of boxes containing moisturized soil. The specimens (30×10 mm) were buried 100 mm beneath the surface of soil, which was regularly moistened with distilled water. At predetermined time points the samples were removed, carefully washed with distilled water in order to ensure the stop of the degradation, dried at room temperature to a constant weight and then were stored in darkness. The specimens were weighed on the PGB200 model analytical balance in order to determine the average weight loss:

$$\text{Weight loss} = \frac{Wo - Wt}{Wo} \times 100$$

where Wo is the initial mass and Wt is the remaining mass at any given time t. All results are the average of three experimental values.

9.2.6 MECHANICAL STRENGTH

Tensile strength, Young's Modulus and Elongation at break were measured according to the ASTM standard method D882-Test method A (ASTM 1997) with application of an Lloyd universal testing machine with a 5 KN capacity at 23 ± 2 °C and $48 \pm 5\%$ RH. Test specimens with a length of 10 cm and a width of 2.5 cm were cut from composite sheets. All specimens were equilibrated in a chamber kept at 18 °C and 35% relative humidity for 24 hr before testing. All these tests were conducted at ambient temperature and an average value of four repeated tests was taken for each material. Specimen used for this test can be observed in the Fig. 9.3.

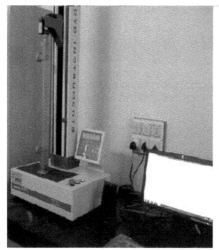

FIGURE 9.3   (a) Universal Testing Machine (UTM) used to measure tensile strength. (b) X-TRAN instrument used for measure the gas barrier properties of the films.

9.2.7   OXYGEN PERMEABILITY TEST

In accordance with ASTM D3985 (ASTM 1995), the oxygen transmission rate (OTR) was determined. The film samples were equilibrated at 22 ± 2 °C and 48 ± 5% RH for at least 48 hrs in a controlled environment cabinet containing a saturated magnesium nitrate solution prior to the analysis. Oxygen permeability (OP) was calculated by the multiplication of the OTR at steady state by the average film thickness divided by the partial pressure difference between the two sides of the film.

9.2.8   WATER VAPOR PERMEABILITY TEST

Based on the ASTM E96–9223 the gravimetric modified cup water method was used to determine water vapor permeability (WVP). Film samples were tested in circular test cups made of polymethylmethacrylate (PMMA). The fan speeds in the cabinets were set at an air velocity of 185 m min^{-1}. The weight loss was monitored until it was certain that water vapor transmission through the film samples had attained a steady state.

9.2.9   SCANNING ELECTRON MICROSCOPY

For the evaluation of the film microstructure scanning electron microscopy (SEM) was used. Before the analysis, the samples were sputter-coated with a thin layer of gold to avoid electrical charging.

9.3   RESULTS AND DISCUSSION

9.3.1   COMPOSITE FILMS FROM MODIFIED CELLULOSE FIBER-REINFORCED WITH PLA/PVA/ POLYPYRROLIDONE

9.3.1.1   MOISTURE ABSORPTION BEHAVIOR

The moisture absorption results are crucial for understanding the performance of cellulose-based composites, since the moisture pickup under immersion in water or exposure to high humidity, intimately relates to such composite properties as mechanical strength, dimensional stability and appearance. Though the PLA, Poly (vinyl alcohol) and Polypyrrolidone has been considered as one of the most promising materials for biodegradable plastics, but because of its poor resistance to water absorption limits its wide applications. Addition of fillers is an effective way of decreasing its sensitivity to moisture and improving mechanical properties. Moisture absorption test was carried for all the composite films in which the modified cellulose and matrix polylactic acid are in the ratio of 10:90, 20:80, 30:70, 40:60, 50:50, 60:40, 70:30, 80:20, 90:10 and 95:05.

It was observed that in all the three modified cellulose as the percentage of modified cellulose increases, moisture absorption decreased. This behavior clearly reflects the presence of hydrophobic moieties onto the fiber surface increase in their resistance towards moisture. Table 9.1 shows the moisture absorption behavior for all the films.

9.3.1.2   BIODEGRADATION IN SOIL

Biodegradation of materials occurs in various steps. Initially, the digestible macromolecules, which join to form a chain, experience a direct enzymatic scission. This is followed by metabolism of the split portions,

TABLE 9.1 Moisture Absorption Studies of Modified Cellulose/(PLA or PVA or PP)) Film Composites

No. of days	%wt. decrease 10:90	%wt. decrease 20:80	%wt. decrease 30:70	%wt. decrease 40:60	%wt. decrease 50:50	%wt. decrease 60:40	%wt. decrease 70:30	%wt. decrease 80:20	%wt. decrease 90:10	%wt. decrease 95:05
2	2.9	2.7	2.7	2.4	1.9	1.7	1.7	1.5	1.4	1.4
4	5.1	4.9	4.8	4.4	3.9	3.6	3.6	3.1	2.9	2.8
6	7.2	6.9	6.7	6.2	5.6	5.1	5.0	4.5	4.3	4.2
8	9.0	8.7	8.4	7.8	7.1	6.5	6.4	5.9	5.6	5.5
10	10.6	10.3	9.7	9.0	8.2	7.5	7.3	6.8	6.4	6.2
12	12.3	12.0	11.0	10.1	9.3	8.6	8.2	7.7	7.2	6.9
14	13.9	13.5	12.2	11.3	10.4	9.6	9.1	8.6	8.0	7.6
16	15.7	15.3	14.5	12.5	11.4	10.7	10.1	9.4	8.7	8.2
18	17.2	16.7	15.8	13.6	12.5	11.8	11.0	10.2	9.5	8.9
20	18.6	18.1	17.1	14.8	13.6	12.9	11.9	11.1	10.2	9.5
22	20.1	19.5	18.3	16.0	14.7	13.9	12.8	12.0	11.0	10.1
24	21.0	20.3	19.5	17.2	15.7	14.9	13.8	12.8	11.7	10.6

MODIFIED CELLULOSE/PLA

TABLE 9.1 (Continued)

No. of days	%wt. decrease 10:90	%wt. decrease 20:80	%wt. decrease 30:70	%wt. decrease 40:60	%wt. decrease 50:50	%wt. decrease 60:40	%wt. decrease 70:30	%wt. decrease 80:20	%wt. decrease 90:10	%wt. decrease 95:05
2	3.0	2.8	2.7	2.4	1.8	1.7	1.6	1.5	1.4	1.4
4	5.3	4.9	4.8	4.5	3.8	3.5	3.5	3.0	2.8	2.7
6	7.1	6.8	6.6	6.1	5.5	5.0	4.9	4.4	4.2	4.1
8	9.0	8.7	8.4	7.8	7.1	6.5	6.4	5.9	5.6	5.5
10	10.7	10.4	9.8	9.1	8.3	7.6	7.4	6.9	6.5	6.3
12	12.1	11.8	10.8	9.9	9.1	8.4	8.0	7.5	7.0	6.7
14	13.8	13.4	12.1	11.2	10.3	9.5	9.0	8.5	7.9	7.5
16	15.7	15.3	14.5	12.5	11.4	10.7	10.1	9.4	8.7	8.2
18	17.4	16.9	16.0	13.8	12.7	12.0	11.2	10.4	9.7	9.1
20	18.5	18.0	17.0	14.7	13.5	12.8	11.8	11.0	10.1	9.4
22	20.3	19.7	18.5	16.2	14.9	14.1	13.0	12.2	11.2	10.3
24	21.3	20.6	19.8	17.5	16.0	15.2	14.1	13.1	12.0	10.9

MODIFIED CELLULOSE/PVA

TABLE 9.1 *(Continued)*

No. of days	%wt. decrease 10:90	%wt. decrease 20:80	%wt. decrease 30:70	%wt. decrease 40:60	%wt. decrease 50:50	%wt. decrease 60:40	%wt. decrease 70:30	%wt. decrease 80:20	%wt. decrease 90:10	%wt. decrease 95:05
2	2.6	2.4	2.4	2.1	1.6					
4	4.7	4.5	4.4	4.0	3.5					
6	6.9	6.6	6.4	5.9	5.3					
8	8.5	8.2	7.9	7.3	6.6					
10	10.2	9.9	9.3	8.7	7.8					
12	11.8	11.5	10.5	9.6	8.8					
14	13.5	13.1	11.8	10.9	10.0					
16	15.2	14.8	14.0	12.1	10.9					
18	16.8	16.3	15.4	13.2	12.1					
20	18.1	17.6	16.6	14.3	13.1					
22	19.5	18.9	17.7	15.4	14.1					
24	20.3	19.6	18.8	16.5	15.0					

MODIFIED CELLULOSE/PP
OSE/PP

leading to a progressive enzymatic dissimilation of the macromolecule from the chain ends. Oxidative cleavage of the macromolecules may occur instead, leading to metabolization of the fragments. Either way, eventually the chain fragments become short enough to be converted by microorganisms.

The studies on biodegradation behavior are important for the application of bio composites in environment. In this work, soil burial experiments were performed for all the ten ratio films. Table 9.2 presents weight loss of various films as a function of biodegradation time.

Above value indicates that weight loss shows an approximately linear relation with degradation time for all the films. The weight decreases for 2 days is average 3–5% and it decreases gradually as the time increase and after 18 days average weight decrease is 16%. The ability of films to degrade depends greatly with physicochemical characteristics of the substrate, such as the degree of crystallinity and polymerization of cellulose, of which the crystallinity degree of cellulose is the most important structural parameters. Crystalline regions are more difficult to degrade. All the ten film composites showed almost same resistance to microorganism attack in the soil. As the microorganism attacks, the composites lose their structural integrity. Undoubtedly, the results obtained to herein reveal that the film composites would not cause any deleterious ecological impact.

9.3.1.3 MECHANICAL PROPERTIES

The mechanical behavior of a polymer can be characterized by its stress-strain properties. This often involves observing the behavior of a polymer as one applies stress to it in order to elongate it to the point where it ruptures.

TABLE 9.2 Biodegradation Studies of Modified Cellulose/(PLA or PVA or PP)) Film Composites

No. of days	%wt. decrease 10:90	%wt. decrease 20:80	%wt. decrease 30:70	%wt. decrease 40:60	%wt. decrease 50:50	%wt. decrease 60:40	%wt. decrease 70:30	%wt. decrease 80:20	%wt. decrease 90:10	%wt. decrease 95:05
2	2.7	2.8	3.0	2.9	2.8	3.1	2.9	3.0	3.3	3.2
4	3.9	3.8	4.1	3.9	3.9	4.2	3.7	4.2	4.2	4.4
6	5.8	5.6	5.9	5.7	5.5	5.6	5.3	5.5	5.7	5.6
8	7.7	7.5	7.6	7.4	7.5	7.4	7.7	7.6	7.8	7.6
10	9.3	9.5	9.4	9.2	9.3	9.6	9.6	9.8	9.7	9.8
12	11.7	11.6	11.7	11.5	11.4	11.6	11.4	11.6	11.7	11.9
14	13.2	13.4	13.6	13.7	13.5	13.8	13.7	13.8	13.9	14.0
16	15.4	15.7	15.6	15.9	15.4	15.6	15.6	15.7	15.5	15.7
18	15.6	15.8	15.7	15.9	15.5	15.6	15.8	15.8	15.6	15.9

MODIFIED CELLULOSE/PLA

TABLE 9.2 (Continued)

	No. of days	%wt. decrease 10:90	%wt. decrease 20:80	%wt. decrease 30:70	%wt. decrease 40:60	%wt. decrease 50:50	%wt. decrease 60:40	%wt. decrease 70:30	%wt. decrease 80:20	%wt. decrease 90:10	%wt. decrease 95:05
MODIFIED CELLULOSE/PVA	2	2.4	2.7	3.1	2.8	2.9	3.1	2.9	3.2	3.4	3.3
	4	3.6	3.8	4.4	3.8	3.9	4.2	3.7	4.2	4.6	4.3
	6	5.6	5.7	5.8	5.7	5.8	5.7	5.5	5.7	5.7	5.7
	8	7.5	7.4	7.6	7.8	7.5	7.7	7.7	7.6	7.9	7.9
	10	9.4	9.7	9.7	9.4	9.8	9.6	9.6	9.8	9.9	9.8
	12	11.9	11.8	11.9	11.9	11.5	11.6	11.4	11.8	11.7	11.8
	14	13.7	13.7	13.6	13.7	13.6	13.8	13.8	13.6	13.8	14.1
	16	15.7	15.9	15.6	15.9	15.7	15.7	15.8	15.7	15.9	15.7
	18	16.6	16.8	16.7	16.9	16.5	16.6	16.8	16.8	16.6	16.9
MODIFIED CEL-LULOSE/PP	2	3.5	3.7	3.4	3.9	3.7					
	4	4.8	4.9	5.2	5.1	4.6					
	6	6.6	6.9	6.8	7.1	6.3					
	8	8.7	8.6	8.5	9.4	9.3					
	10	11.2	11.4	11.3	11.5	11.4					

TABLE 9.2 *(Continued)*

No. of days	%wt. decrease 10:90	%wt. decrease 20:80	%wt. decrease 30:70	%wt. decrease 40:60	%wt. decrease 50:50	%wt. decrease 60:40	%wt. decrease 70:30	%wt. decrease 80:20	%wt. decrease 90:10	%wt. decrease 95:05
12	13.5	13.6	14.4	14.6	14.8					
14	14.7	15.1	15.3	15.4	15.6					
16	16.6	16.9	16.4	16.8	16.5					
18	16.8	17.2	16.6	16.8	16.6					

TABLE 9.3   Typical Tensile Properties of Modified Cellulose/(PLA or PVA or PP) Film Composites

	Modified cellulose+ PLA or PVA or PP	Tensile Strength (MPa)	Young's Modulus (MPa)	Elongation at break (mm)	Graphs are 1. Tensile strength versus Parts per hundred Modified Cellulose 2. Young's Modulus versus Parts per hundred Modified Cellulose 3. Elongation at break versus Parts per hundred Modified Cellulose
MODIFIED CELLULOSE/PLA	10+90	22.41	827.76	302.02	
	20+80	22.77	870.44	272.09	
	30+70	22.79	880.85	262.41	
	40+60	25.66	970.32	211.95	
	50+50	38.44	1342.92	166.79	
	60+40	38.51	1510.09	151.45	
	70+30	39.53	1800.78	107.28	
	80+20	39.73	3184.81	96.61	
	90+10	40.15	4361.22	39.44	
	95+05	41.30	4947.64	33.76	
MODIFIED CELLULOSE/PVA	10+90	18.36	720.75	275.76	
	20+80	19.00	742.09	274.97	
	30+70	19.56	759.47	272.18	
	40+60	19.62	789.88	249.26	
	50+50	19.97	807.53	218.33	
	60+40	20.12	851.60	88.56	
	70+30	20.54	875.45	71.98	
	80+20	21.16	893.87	70.35	
	90+10	21.23	1093.20	58.04	
	95+05	22.41	1316.18	41.07	

Modified cellulose+ PLA or PVA or PP	Tensile Strength (MPa)	Young's Modulus (MPa)	Elongation at break (mm)	Graphs are 1. Tensile strength versus Parts per hundred Modified Cellulose 2. Young's Modulus versus Parts per hundred Modified Cellulose 3. Elongation at break versus Parts per hundred Modified Cellulose		
10:90	13.42	655.70	333.15	
30:70	15.21	866.23	297.44	
50:50	18.33	1276.53	196.79	
70:30	19.66	1432.75	132.26	
90:10	21.88	1543.88	67.44	

MODIFIED CELLULOSE/PP

Mechanical properties of all three types of modifies cellulose (with PLA, PVA, PP) are studied in present work. An increasing trend in tensile strength and Young's modulus with modified cellulose fiber content is found from (Table 9.3). It was observed that tensile strength and Young's modulus of the films increases as the percentage composition of the modified cellulose increases. This enhancement indicates the effectiveness of the modified cellulose as reinforcement. With the increasing of cellulose content, the interactions between the cellulose and the matrix is improved and crack propagation was inhibited, which resulted in the increased tensile strength and Young's modulus. Contrarily, it illustrated that there was interfacial adhesion between cellulose and the matrix; otherwise, it would result in premature composite failure because the reinforcing cellulose simply pulled out of the matrix without contributing to the strength or stiffness of the material. It was observed that, there is a sudden increase of tensile strength at 50:50 compositions, which indicates that good compatibility of the matrix and the fiber at that composition.

9.3.1.4   OXYGEN PERMEABILITY TEST

Oxygen permeability depends on chain flexibility, phase and physical state of the polymer and packing of its molecules. The most permeable polymers are amorphous, with very flexible chains, in a high elastic state. The gas permeability of crystalline polymer is much lower. The high molecular weight glassy polymers with rigid chains have very low gas permeability. With decreasing chain flexibility gas permeability decreases. Closer packing of the molecules supports permeability resistance.

TABLE 9.4   OTR Values of Ten (Modified Cellulose/(PLA or PVA or PP)) Film Composites

	Modified cellulose+ PLA	OTR values; Cc/sqm/day/atm	Graph for Oxygen permeation versus Parts per hundred Modified Cellulose in a (modified cellulose/(PLA or PVA or PP) blend
MODIFIED CELLULOSE/PLA	10+90	1876	
	20+80	1704	
	30+70	1543	
	40+60	1443	
	50+50	1342	
	60+40	1231	
	70+30	1123	
	80+20	1086	
	90+10	989	
	95+05	843	

MODIFIED CELLULOSE/PVA	10+90	2005
	20+80	1907
	30+70	1854
	40+60	1701
	50+50	1589
	60+40	1498
	70+30	1245
	80+20	1087
	90+10	987
	95+05	843
MODIFIED CELLULOSE /PP	10:90	2654
	30:70	2324
	50:50	1456
	70:30	1213
	90:10	877

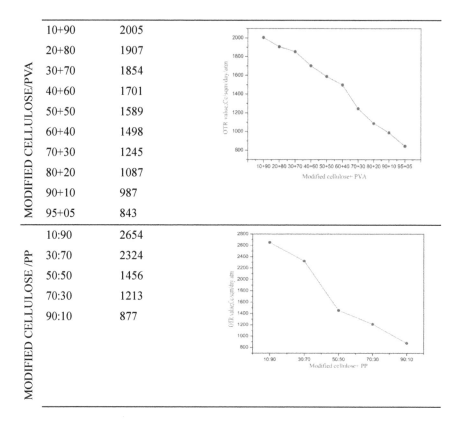

Table 9.4 represents OTR (Oxygen Transmission Rate) values for all the ten ratio films. Generally, hydrophilic polymeric films have shown good oxygen barrier property. As it can be observed there was an improvement in oxygen barrier properties of the films as the percentage of modified cellulose increases. It was observed that there is a great decrease in oxygen transmission rate as the percentage composition of the modified cellulose increases. It is obvious that modified cellulose played a powerful role in improving the oxygen gas barrier properties. The increased molecular interaction resulted in a film with compact structure and low OTR value. Oxygen Transmission Rate increases as the percentage of modified cellulose decreases because intermolecular bonding between fiber and matrix decreases. This resulted in a phase separation among the main components where the film could not be formed well, facilitating the oxygen permeation. So it was more advantageous to improving the gas barrier properties by increasing the percentage of modified cellulose. This result

indicates the potential of these films to be used as a natural packaging to protect food from oxidation reactions.

9.3.1.5   WATER VAPOR PERMEABILITY TEST

The water vapor permeability of films depends on many factors, such as the integrity of the film, the hydrophilic-hydrophobic ratio, the ratio between crystalline and amorphous zones and the polymeric chain mobility.

Table 9.5 represents Water Vapor Transmission Rate values for all the ten ratio films. It was observed that there is a small decrease in water vapor transmission rate as the percentage composition of the modified cellulose increases. This is because as the percentage composition of modified cellulose increases, hydrophilicity of the film decreases. This phenomenon could be related to the significant hydrogen bonding interaction with water. The comparison between OTR and WVTR indicates that modified cellulose is greatly effective in obstructing the oxygen permeation, but less effective in retarding the water vapor permeation. This results shows that these films may impede moisture transfer between the surrounding atmosphere and food, or between two components of a heterogeneous food product. This property is very much use full in packaging application.

TABLE 9.5   WVTR Values of Ten (modified cellulose/(PLA or PVA or PP)) Film Composites

	Modified cellulose+ PLA	WVTR values; gm/sqm/day	Graph for Water vapor permeation versus Parts per hundred Modified Cellulose in a (modified cellulose/PLA) blend
MODIFIED CELLULOSE/PLA	10+90	6.752	
	20+80	5.985	
	30+70	5.128	
	40+60	4.540	
	50+50	4.121	
	60+40	3.769	
	70+30	3.088	
	80+20	2.789	
	90+10	2.075	
	95+05	1.121	

TABLE 9.5 *(Continued)*

	Modified cellulose+ PLA	WVTR values; gm/sqm/day	Graph for Water vapor permeation versus Parts per hundred Modified Cellulose in a (modified cellulose/PLA) blend
MODIFIED CELLULOSE/PVA	10+90	7.312	
	20+80	6.976	
	30+70	6.123	
	40+60	5.554	
	50+50	4.986	
	60+40	4.321	
	70+30	3.897	
	80+20	3.123	
	90+10	2.786	
	95+05	1.876	
MODIFIED CELLULOSE/PP	10:90	6.212	
	30:70	5.965	
	50:50	4.167	
	70:30	3.431	
	90:10	2.543	

9.3.1.6 MORPHOLOGICAL OBSERVATION

SEM micrographs [10] were taken for four film composites in different magnification. Figure 9.4 shows 10:90 (modified cellulose/(PLA or PVA or PP)) ratio film composite and clearly shows that the interfacial adhesion between the cellulose fiber and matrix was slightly poor. Moreover, the presence of cellulose fiber aggregates, some of them visible with the naked eyes, provided strong evidence for the nonhomogeneity of the material, 50:50 ratio film composite and provided good interfacial adhesion between two components, 70:30 ratio film composite and 90:10 ratio film composite, showed a more uniform morphology, that is, the fibers were less visible because they were buried more effectively in the matrix. However, the fiber pullout length is not large, indicating a good fiber-matrix bonding. In this connection, it can be argued that there is an improvement of interfacial strength in the film composite as the percentage composition of modified cellulose increases. In other words bulk fiber reinforcing film composites have the most uniform distribution of fiber and the most compatible interface in film composite.

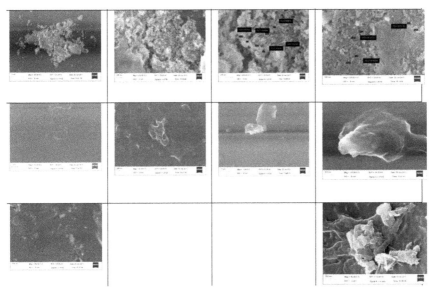

FIGURE 9.4   Scanning electron micrograph of modified cellulose/(PLA or PVA or PP) blend at different ratio.

9.4   CONCLUSIONS

1. Modification of cellulose by using 2-(Trifluromethyl) benzoylchloride has been found to be key investigation in this work.
2. Biodegradable composite films were developed by film casting method using modified cellulose Poly (vinyl alcohol); PLA; Polypyrrolidone in different compositions and hybrid biodegradable composite films were developed by film casting method using modified cellulose with Poly(vinyl alcohol) and Poly(lactic acid); modified cellulose with Poly(vinyl alcohol) and Polypyrrolidone in different compositions.

3. The film composites were characterized by mechanical, moisture absorption, oxygen permeability, water vapor permeability and biodegradable properties.

4. Film composites produced by this method shows very good bio-degradation behavior, which renders them advantageous in terms of environmental protection.

5. Moisture absorption result shows that modified cellulose plays a great role in increasing the composite properties such as mechanical properties, since as the proportion of modified cellulose increases water uptake by the film composite was less.

6. OTR and WVTR test values shows that modified cellulose plays a powerful role in increasing the gas barrier properties. Hence these films can be used as a packaging to protect food from oxidation reaction and moisture.

7. The produced film composites possess higher tensile strength and Young's Modulus as the proportion of modified cellulose increases. So modified cellulose plays vital role in increasing the mechanical strength of film composites.

KEYWORDS

- 2-(trifluromethyl)benzoylchloride
- Cellulose
- Films modified
- Mechanical properties
- Oxygen permeability
- Polypyrrolidone
- Polyvinyl alcohol
- Water vapor permeability

REFERENCES

1. Petersen, K., Nielsen, P. V., Bertelsen, G., Lawther, M., Olsen, M. B., Nilssonk, N. H., & Mortensen, G. (1999). Potential of Biobased Materials for Food Packaging, Trends Food Sci Tech, 10, 52–68.

2. Tserki, V., Matzinos, P., Zafeiropoulos, N. E., & Panayiotou, C. (2006). Development of Biodegradable Composites with Treated and Compatibilized Lignocellulosic Fibers, J Appl Polym Sci., 100, 4703–4710.

3. Soykeabkaew, N., Supaphol, P., & Rujiravanit R. (2004). Preparation and Characterization of Jute-and Flax-reinforced Starch-Based Composite Foams, Carbohyd Polym, 58, 53–63.

4. Sreekala, M. S., Goda, K., Devi, P. V. (2008). Sorption Characteristics Of Water, Oil And Diesel In Cellulose Nanofiber Reinforced Corn Starch Resin/Ramie Fabric Composites. Compos Interface, 15, 281–299.

5. Averous, L., Fringant, C., & Moro, L. (2001). Plasticized Starch-cellulose Interactions in Polysaccharide Composites, Polymer, 42, 6565–6572.

6. Sandeep, S. L., Viveka, S., Madhu Kumar, D. J., Dinesha Bhajanthri, R. F., & Nagaraja, G. K. (2012). *Preparation and Characterization of Modified Cellulose Fiber-Reinforced Polyvinyl Alcohol/Polypyrrolidone Hybrid Film Composites.* J Macromol Sci A, 49, 639-647.

7. Sandeep, S. L., Madhu Kumar, D. J., Viveka, S., & Nagaraja, G. K. (2012). Preparation and Properties of Biodegradable Film Composites Using Modified Cellulose Fiber-Reinforced with PVA. ISRN Polymer Science Article ID, 154314, 8p.

8. Son, S. J., Lee, Y. M., Im, S. S. (2000). Transcrystalline Morphology And Mechanical Properties In Polypropylene Composites Containing Cellulose Treated With Sodium Hydroxide And Cellulase. J Mater Sci., 35, 5767–5778.

9. Sandeep, S. L., Viveka, S., Madhu Kumar, D. J., & Nagaraja, G. K. (2012). Preparation and Characterization of Modified Cellulose Fiber-Reinforced Polypyrrolidone Film Composites. Inventi Rapid: Pharm Tech ISSN 0976-3783.

10. Sandeep, S. L., Viveka, S., Madhu Kumar, D. J., & Nagaraja, G. K. (2012). Preparation and Properties of Composite Films from Modified Cellulose Fiber-reinforced with PLA. Der Pharma Chemica, 4, 159–168.

CHAPTER 10

NATURAL BIORESOURCES: THE UNENDING SOURCE OF NANOFACTORY

BALAPRASAD ANKAMWAR

CONTENTS

Abstract 196
10.1   Introduction 196
10.2   Biosynthesis of Nanomaterials 197
10.3   Conclusions and Future View 199
Keywords 200
References 200

ABSTRACT

Natural resources are indispensable for the life cycle. In this article few number of natural bio-resources exploited for synthesis of nanoparticles are discussed in brief. The large pool of basic raw materials of reducing and capping agents required for the synthesis nanomaterials could be used as the ecofriendly and body benign method of nanomaterials synthesis. It has various applications with main focus on biomedical applications. This article elaborates the use of microorganisms, various parts of plants such as leaves, fruits and herbs for the synthesis of nanomaterials with main focus on gold and silver nanoparticles. Nature is using nanoscale everywhere in living and nonliving things.

10.1   INTRODUCTION

Nanoscience and nanotechnology (NT) has started in the nature, since time immemorial. Modern analytical tools such as TEM, SEM, AFM, STM and XRD, etc., could explore and support this technology. As a result NT is expanding its horizon in the modern life, like Information Technology (IT), Biotechnology (BT). Ultimately, the type of various sources of nanomaterials synthesis became more significant. Natural bioresources became unending source of nanofactory in this regard, since it has large pool of reducing and capping agents required for nanomaterials synthesis. Biological, synthesis of triangular gold nanoprisms using natural reducing and capping agents from extracts of *Cymbopogon flexuous* [1], *Tamarindus indica* [2], silver nanobuns [3] using *Murraya Koenigi* leaf and gold and silver nanoparticles using *Emblica officinalis* [4] fruit and *Terminalia catappa* [5] are few examples of exploiting natural resources for nanofactory. Gold nanoprisms could be exploited for hyperthermia of tumors [6] and silver nanoparticles in catalysis [7], etc. Red bioluminescent of Jellyfish, various beautiful colors of butterfly, exterior of toucan breaks, flies stick to walls [8] using gift of natural NT. Currently, there is growing need to develop ecofriendly and body benign nanoparticles synthesis processes in the synthesis protocols by mimicking nature to avoid adverse effects especially in the biomedical and other similar kind of applications. Obviously, researchers in this field diverted their attention towards the exploiting natural resources for the synthesis of biocompatible nanostructures. Natural bio resources are derived from the environment. Some of them are essential for our survival while most are used for satisfying our wants. There should be a perfect balance between them. Every man-made product is composed of natural resources. At most we can only manipulate natural resources for satisfying our wants. Here briefly, the uses of natural bio-resources for the synthesis of nanostructures are discussed.

10.2   BIOSYNTHESIS OF NANOMATERIALS

10.2.1   USE OF MICROORGANISMS

Large pool of biological species attracted researchers towards the use of inorganic molecules for the synthesis of inorganic nanocrystals. As a result extensive work has been carried out using microorganisms for biosynthesis of nanoparticles. Here, some of the organisms exploited as the source of nanofactory for the synthesis of nanoparticles are discussed in brief. Gold and silver nanoparticles are most studied using biosynthesis route. The most important reason is their high reduction potentials and applications are limitless. Moreover, synthesis of gold and silver nanoparticles in various size and shapes using different permutations and combinations of precursor metal salt, reducing and capping agent has been well proven. In addition to it, most easy method of synthesis has attracted huge number of researchers around the globe. Despite this fact various attempts are also being made for biosynthesis of other kind of nanoparticles. Various microbial species from mines of metal ores inspired the biosynthesis of concern metal nanoparticles using microorganisms, such as Pseudomonas stutzeri AG259, isolated from silver mines, has been shown to produce silver nanoparticles [9]. Kazem Kashefi and Adam Brown fed the *Cupriavidus metallidurans* unprecedented amounts of gold chloride and reported that the metal-tolerant bacteria can grow on massive concentrations of gold chloride or liquid gold and in about a week; the bacteria transformed the toxins and produced a 24-karat gold nugget. Michigan State University researchers have discovered a bacterium's ability to withstand incredible amounts of toxicity is key to creating 24-karat gold [10]. Recently effective strategy to prepare hexagonal phase CdS nanocrystals micrometer-long hybrid nanofibers by deposition onto the substrate of hydrated bacterial cellulose nanofibers were achieved via a simple hydrothermal reaction between $CdCl_2$ and thiourea at relatively low temperature and these hybrid nanofibers demonstrated high-efficiency photocatalysis with 82% methyl orange degradation after 90 min irradiation and good recyclability [11]. Magnetotactic bacteria exploits magnetosomes to orient themselves in responses to external magnetic fields, including the earth's magnetic field [12] and different types of bacteria make magnetite crystals with a variety of morphologies using one or more magnetosome chains [13].

10.2.2   USE OF PLANT AND HERB EXTRACTS

The Lemon grass (*Cymbopogon flexuous*) plant extract when reacted with aqueous Au^{3+} ions yields a high percentage of thin, flat, single crystalline gold nanotriangles in single step [1]. The sharp vertices of gold nanotriangles can be used as Scanning Tunneling Microscopy (STM) conductive tips and its exciting application based on the large Near Infra Red (NIR) absorption of the particles could be in hyperthermia of tumors [6]. The size and thickness of the nanotriangles being in the range 200–500 nm and 8–18 nm, respectively makes manipulation onto cantilevers feasible. Another method of biosynthesis of gold nanotriangles using *Tamarindus indica* with size and thickness in the range of 200–500 and 20–40 nm, respectively, could be exploited as methanol sensor [2] in addition to hyperthermia of tumors. Defective silver nanobuns with diameter in the range of 20–40 nm using *Murraya Koenigi* leaf extract [3] could be exploited in SERS studies. The silver and gold nanoparticles synthesized using *Emblica officinalis* fruit extract indicated that they ranged in size from 10 to 20 nm and 15 to 25 nm, respectively [4]. The reduction of the metal ions and the stabilization of the Ag and Au nanoparticles were believed to occur by the various hydrolysable tannins present in the EO fruit [4]. Whereas gold nanoparticles obtained in the diameter range of 10–35 nm by *Terminalia catappa* leaf extract was extremely stable due to the antibacterial [14] and antioxidant [15] properties of biomolecules present in the *Terminalia catappa* leaf extract have facilitated excellent stability of the nanoparticles. Wang et al. describes the *Barbated Skullcup* herb extract-mediated biosynthesis of gold nanoparticles and its primary application in electrochemistry [16]. They [16] reported the extracellular synthesis of gold nanoparticles ranged in size 5–30 nm using *Barbated*

Skullcup (BS) herb (a dried whole plant of *Scutellaria barbata* D. Don as the reducing agent.

10.2.3 NANO IN NATURE

There are innumerable examples exploiting nanoscale for practical purposes, such as the nanometer-sized biodegradable threads of spider silk are stronger, by weight, than high-tensile steel. It is also elastic enough to stretch up to 10 times its initial length [8]. Hippo sweat contains compounds that absorb light in the range of 200–600 nanometers. This compound protects the hippo's skin like sunscreen [17]. A jellyfish-type invertebrate, called a siphonophore, uses red bioluminescent lures created at the nanoscale to attract prey, since red isn't easily visible underwater [18]. The colors of beetle and butterfly wings come from the scattering of light. Light hits the nanostructures on their scales. These nanostructures are typically smaller than the wavelengths of visible light (smaller than 400 nanometers) [19]. Another reason for color is the interference of different wavelengths of light [20]. Beetles and flies also have nanostructures that help them stick to walls, ceilings and what appear to be smooth surfaces [21].

These are very few representative examples, where nanoscale has been used extensively for the survival of living things; of course, it is equally applicable to nonliving things too. The matter of serious concern is we should use natural bio-resources for the synthesis of nanomaterials without disturbing environment and natural cycle before it will be too late.

10.3 CONCLUSIONS AND FUTURE VIEW

In this article few representative biosynthesis methods of nanomaterials are discussed, however, it is unending source of the natural bio-resources of reducing and capping agents for the synthesis of nanoparticles are concern. Here few bacterial sources and plants and herb extracts are discussed in brief. Bio-nanotechnology is steel under infancy. It is new emerging area and could be better future prospective. Various composition, size and shaped nanoparticles can be exploited for different applications. Nature is exploiting nanoscale for living and nonliving things since time immemorial. Though large quantum of work has been reported as of today in this regard, still huge scope is waiting for further advancement and cutting age research in this area.

KEYWORDS

- **Biosynthesis**
- **Nanoparticles**
- **Reducing agents**
- **Green synthesis**
- **Biological reduction**
- **Biological capping agents**

REFERENCES

1. Shiv Shankar, S., Rai, A., Ankamwar, B., Singh, A., Ahmad, A., & Sastry, M. (2004). Biological Synthesis of Triangular Gold Nanoprisms, Nature Materials, 3, 482–488.
2. Ankamwar, B., Chaudhary, M., & Sastry, M. (2005). Gold nano-triangles biologically synthesized using tamarind leaf extract and potential application in vapor sensing. Syn React Inorg Metal Org. Nano-Metal Chem., 35, 19–26.
3. Ankamwar, B., Mandal, G., Sur, U. K., & Ganguly, T. (2012). An effective biogenic protocol for room temperature one-step synthesis of defective nanocrystalline silver nanobuns using leaf extract, Digest J. Nano. Biostruct, 7, 599–605.
4. Ankamwar, B., Damle, C., Ahmad, A., & Sastry, S. (2005). Biosynthesis of gold and silver nanoparticles using Emblica officinalis fruit extract, their phase transfer and transmetallation in an organic solution. J. Nanosci Nanotech., 5, 5, 1665–1671.
5. Ankamwar, B. (2010). Biosynthesis of gold nanoparticles (Green-gold) using leaf extract of Terminalia catappa. E-J Chem., 7, 1334–1339.
6. Hirsch, L. R., Stafford, R. J., Bankson, J. A., Sershen, S. R., Rivera, B., Price, R. W., Hazle, J. D., Halas, N. J., & West, J. L. (2003). Nanoshell mediated near infrared thermal therapy of tumors under magnetic resonance guidance, Proc Natl Acad. Sci USA, 100, 13549–13554.
7. Zhou, Q., Qian, G., Li, Y., Zhao, G., Chao Y., & Zheng, J. (2008).Two dimensional assembly of silver nanoparticles for catalytic reduction of 4-nitroaniline. Thin Solid Films, 516, 953–956.
8. http://shasta.mpi-stuttgart.mpg.de/research/Bio-tribology.htm.
9. Ramanathan, R., O'Mullane, A. P., Parikh, R. Y., Smooker, P. M., Bhargava, S. K., & Bansal, V. (2011). Bacterial kinetics controlled shape-directed biosynthesis of silver nanoplates using morganella psychrotolerans, Langmuir, 27, 714–719.
10. http://www.sciencedaily.com/releases/2012/10/121002150031.htm.

11. Yang, J., Yu, J., Fan, J., Sun, D., Tang, W., & Yang, X. (2011). Biotemplated preparation of CdS nanoparticles/bacterial cellulose hybrid nanofibers for photocatalysis application, J. Hazard Mater, 189, 377–83.

12. Blakemore, R. (1975). Magnetotactic Bacteria Science, 190, 377–379.

13. Baeuerlein, E. (Ed.). (2005). Biomineralization: progress in biology, molecular biology, and application; Wiley–VCH Weinheim, 44, 4833–4834.

14. Pawar, S. P., & Pal, S. C. (2002). Antimicrobial activity of extracts of Terminalia catappa root, Indian J Med Sci., 56, 276–278.

15. Ko, T. F., Weng, Y. M., & Chiou, R. Y. (2002). Antimutagenicity of Supercritical CO2 Extracts of Terminalia catappa Leaves and Cytotoxicity of the Extracts to Human Hepatoma Cells. J. Agric. Food Chem., 11, 5343–5348.

16. Wang, Y., He, X., Wang, K., Zhang, X., & Tan, W. (2009). Barbated Skull Cup Herb extract-mediated biosynthesis of gold nano-particles and its primary application in electrochemistry, Colloids and Surfaces B: Biointerfaces, 73, 75–79.

17. http://www.pbs.org/kratts/world/africa/hippo/index.html.

18. http://www.coml.org/medres/high2005/highlightimages.htm.

19. http://pubs.acs.org/cen/critter/butterfly.html.

20. http://acept.la.asu.edu/PiN/rdg/interfere/interfere.shtml.

21. http://shasta.mpi-stuttgart.mpg.de/research/Bio-tribology.htm.

INDEX

A

Acceleration voltage, 105, 106, 108
Acetobacter xylinum bacteria, 32
Acute neonatal diarrhea, 53
Adventitia's primary function, 17
  body system, 17
    nerve networks, 17
    vascular, 17
Agro waste, 133
Alginate, 146, 147, 156
  L-guluronic acid, 156
Alteromonas macleodii, 157
  Deepsane, 157
Amidated Linseed polyol, 12
Angiogenesis, 152
  matrix metalloprotease inhibitors, 152
Animal reservoirs, 55
  beavers, 55
  cats, 55
  cattle, 55
  dogs, 55
Anionic surfactant, 80, 103
  amphiphilic nature, 80
  micelles, 80
  pyrrole, 80
Anticancer activities, 152
  apoptosis of carcinogenic cells, 152
  inhibiting metastasis, 152
Anticoagulant properties, 34
Antithrombotic agents, 40
Apicomplexan (oo)cysts membrane protein, 62
Aqueous solution, 35, 76, 78, 80, 93
Aromatic functionalities, 65
  low cell adhesion, 65
Arterial environment, 20
Arterial wall structure, 16
  adventitia, 16
  intima, 16
  media, 16
Artificial blood vessel, 19
Atala's group, 20
  bovine ECs (bECs), 20
  bovine SMCs (bSMCs), 20
Atlantic Innovation Fund, 94
Augur group, 18
Autologous grafts, 16

B

Bacterial, 11, 21, 29, 64, 146, 151, 158, 160
  cell serves, 146
  cellulose, 32
  vascular grafts, 34
Bead-free quality, 18
Biodegradable polyester polymers, 17
  copolymer form of a poly D, 17
  L lactic-coglycolic acid (PLGA), 17
  poly glycolic acid (PGA), 17
  poly lactic acid (PLA), 17
  polycaprolactone (PCL), 17
Bio-hybrid material, 41
Bio-hybrid scaffolds, 17, 19, 40
Biological capping agents, 200
Biological reduction, 200
Biomechanical properties, 23, 37, 39, 40
  burst pressure, 39
  cyclic fatigue, 39
  degradation and growth enhancing characteristics, 39
  elastic of modulus, 39
  suture retention, 39
  tensile strength, 39
Biomedical applications, 19, 196
Biopolymers, 16, 19, 24, 26, 30, 35, 39–41, 60, 61, 66, 67, 147, 154, 170
Bioprospecting, 160
Biosynthesis, 159, 197–200

Biotechnological applications, 147
  antioxidative, 147
  antithrombotic, 147
  antiangiogenic, 147
  antimetastatic, 147
  immunoproliferative, 148
  antiviral, 148
  wound healing properties, 148
Blood coagulation, 30, 153
  cellular and matrix interaction, 30
  fibrin, 30
  fibrinogen, 30
  fibrinolysis, 30
  inflammation, 30
  neoplasia, 30
  wound healing, 30
Blood vessel failure, 16
  atherosclerosis, 16
Blood vessel matrix-structure, 19
Blood-clotting scenario, 30
Bone marrow stem cell (BMSC), 155
Botanically turmeric, 125
  Curcuma Longa, 125

C

Cadaver tissues, 19
Carbohydrate polymers, 146, 160
Carbon coated copper grid, 4
Cardiovascular disease (CVD), 16
Cardiovascular surgery, 38
Carotid artery, 31, 33
  pigs, 33
  rats, 33
Cartilaginous applications, 32
Cartridge filter, 57
Cell-binding sites, 27
Cell-encapsulation, 30
Cell-interaction characteristics, 19
Cell-scaffold interactions, 24
Cellular adhesion, 64, 65
Cellular physiology, 16
  intima aids, 16
    prevention of thrombosis, 16
    infection, 16
    inflammation of the blood vessel, 16
    underlying tissue, 16

Cellular proliferation, 17
Cellular recognition, 18
Cellulose Nanocrystal, 102–104, 116, 117
  US treatment, 102
Cellulose, 32
  versatile biopolymer, 32
    biocompatibility, 32
    biodegradability, 32
    hydrophilicity, 32
    mechanical tenability, 32
    moldability, 32
Center for Nanoscale Materials and Bioin-
  tegration, 40
Chelating functional groups, 149
  phosphates, 149
  sulfates, 149
Chemical oxidative polymerization, 74, 76
  physical properties, 76
    morphology, 76
    conductivity, 76
    environmental stability, 76
    optical properties, 76
Chemical polymerization, 74–76, 78
Chemically treated turmeric petiole fiber
  (TPF CT), 128
Chemically treated turmeric stem fiber
  (TSF CT), 128
Chlorine treatment, 54
Chronic inflammatory, 39
Circumferential testing, 34
Clear filaments, 126
Clinical infection, 53
Coating properties, 3
  corrosion resistance, 3
  flexibility, 3
  hardness, 3
  impact resistance, 3
  mechanical properties, 3
Coelectrospun fibers, 28
Collagen, 19
  cytotoxic response, 19
  hematological properties, 19
  low antigenicity, 19
  low inflammatory, 19
Colonic microflora, 156
  enzyme dextranase, 156

Commercial applications, 146
Composite, 2, 20, 31, 37, 74, 77–94
Conduction band (CB), 89
Core-shell structure, 74, 80, 93, 94
Coronary artery bypass graft (CABG)
    surgery, 16
Coronary artery bypass operations, 16
Cosmeceutical market, 157
Cosmeceutics, 148, 157
    antiaging, 148
    moisturizing, 148
    skin protection, 148
Critical control points, 56
Cross-linking agent, 27
Cryptosporidium adhesion, 51
    biofilms, 51
    clays, 51
    metal oxides, 51
    natural organic matter, 51
    quartz, 51
    silanes, 51
Cryptosporidium oocytes, 50, 57, 60, 62
    thin walled, 62
    thick walled, 62
Culture media, 24
Culture medium, 32
Cumbersome processes, 3
Cutaneous wound healing, 30
    cellular infiltration, 30
    tissue formation, 30
Cyst stage, 51
Cyst surface proteins, 62, 65, 66
Cytoskeletal spreading, 30

D

Decellularized tissue, 18
Derjaguin, Landau, Verwey and Overbeek
    (DLVO) theory, 55
    electrostatic double-layer forces, 65
    short-range attractive Lifshitzvander
    Waals forces, 65
Desoxyribonucleic acid (DNA), 59
Detection, 50, 53, 57–59, 67
Dextran, 30, 146, 156, 159, 160
Diarrheal disease, 50

Differential scanning calorimetry, 79, 102,
    105
Distinct matrix/cell layer, 17
Disulfide-bonded globular structure, 62
Divalent cations, 66
Diverse applications, 60
Donor site morbidity, 16
Dual lamellae population, 109

E

ECM-like formation, 18
EDS image, 80
Elastic mechanical properties, 17
Electrical double layer, 56
    charged groups, 56
        amines, 56
        ionized organic acids, 56
Electro spun biomimetic scaffolds, 18
Electropositive charge, 56, 63
Electrospinning, 18, 20, 24–26, 32, 35–41
Electrospinning technique, 20, 37, 38
    multilayered scaffold, 38
        elastin with tunable mechanical
        properties, 38
        gelatin, 38
        polyglyconate, 38
Electrostatic force, 18
Electrosteric interactions, 63
Elemental analysis, 79
    X-ray spectroscopy, 79
Endothelium cell layer, 39
Erlenmeyer flask, 4
Evaluation of characteristics, 66
    polymer composition, 66
    surface roughness, 66
    wettability, 66
Exopolysaccharides, 146, 151, 152, 160
Extracellular biopolymers, 60
Extracellular environment, 146

F

Fabricated specimens, 124, 127
    flexural, 124
    impact strength properties, 124
    specific flexural, 124

specific tensile, 124
Factors controlling adhesion, 66
Fecal pollution, 50
Fiber diameters, 24
Fiber surface treatments, 172
    benzylation, 172
    peroxide treatment, 172
    silane treatment, 172
Fiber volume fraction, 128, 130, 132
    flexural strength, 130
    modulus, 130
    specific flexural strength, 130
Fibrillar extracellular matrix, 63
    double inner membrane, 63
    outer filamentous wall, 63
Films modified, 192
Fluffy glycoprotein layer, 55
Fourier-transform infrared (FTIR), 74
Free radicals, 150
    degenerative diseases, 150
        asthma, 150
        atherosclerosis, 150
        cancer, 150
        degenerative eye, 150
        diabetes, 150
        inflammatory joint disease, 150
        senile dementia, 150
Freeze-drying step, 104
Freeze-drying technique, 37
Functional groups, 2, 18, 50, 63, 64, 67, 79, 85, 146, 147, 149, 158, 171
    active methylenes, 2
    double bonds, 2
    hydroxyls, 2
    oxiranes, 2

G

Gelatin, 19, 24, 26–30, 37–40
    biocompatibility, 26
    biodegradability, 26
    biological origin, 26
Gellan gum, 155
Giardia, 50, 51, 54–61, 63, 66, 67
Global health, 16
    medical technologies, 16

surgical, 16
Global issue, 50
Glycol functionalities act, 64
    preventative manner, 64
Glycoprotein layer, 66
Gold nanoparticles, 136, 142, 198
Gold standard method, 39
    revascularization procedures, 39
Granular medium, 55
    filter, 55
Green Chemistry, 2, 3, 12
Green resources, 2
    cellulose, 2
    chitosan, 2
    starch, 2
Green synthesis, 66, 200
Greener hybrid materials, 2
    coatings, 2
    paints, 2
Ground water sources, 57

H

Halomonas Maura, 156
    galactose, 156
    glucose, 156
    glucuronic acid with functional sulfate, 156
    mannose, 156
    phosphate groups, 156
Hand lay-up technique, 126
Healing skin wounds, 154
Health-promoting additives, 157
Healthy graft veins, 16
Heavy metal sequestration, 148
Hollow glass tubes, 18
Homo-crystal form, 106
Homogenous solutions, 8
Host of monomers/polymers, 2
    diols, 2
    epoxies, 2
    polyester amides, 2
    polyesters, 2
    polyols, 2
    polyurethanes, 2
Human aortic endothelial cells (HAEC), 35

Human clinical cases, 53
Human coronary artery endothelial cells
    (HCAECs), 22
Human cryptosporidiosis, 52
Human fibroblasts, 18
Human pathogenic species, 50, 53
Human tropoelastin, 24, 40
Human umbilical vein endothelial cells
    (HUVECS), 31
Human umbilical vein smooth muscle cells
    (HUVSMCs), 35
Hydrolysis-condensation reaction, 3
Hydrophobic polymeric matrices, 103
Hydrothermal origin, 157
Hydroxylation procedure, 4
    glacial acetic acid, 4
    hydrogen peroxide, 4
    sulfuric acid, 4

I

Ideal bio-hybrid vascular graft, 40
Immune histo chemical labeling, 33
Immuno electron microscopy, 62
Immunomagnetic separation (IMS), 57
Information collection rule, 57
Innermost layer (intima), 16
    vessel lumen, 16
        collagen type IV, 16
        elastin, 16
        monolayer of endothelial cells
            (ECs), 16
Intact biopolymer, 146
Intercellular Adhesion Molecule, 157
    Keratinocytes, 157
Interconnecting pore-networks, 18
Internal trophozoites, 55
Interpenetrating polymer network (IPN),
    30
Intestinal parasite ova stages, 55

J

JEOL JSM-5600 SEM, 79
Jeol JEM-2000EX, 105

K

Kaplan's group, 35

Keratinocyte proliferation, 157
    hyper-reactive, 157
    protect irritable, 157
    sensitive skin from inflammation, 157
Kirby Bauer method, 10
    disk diffusion assay, 10
        Pseudomonas aeruginosa, 10
        Staphylococcus aureus, 10
Krypton ion laser, 140

L

Lactic acid bacteria (LAB), 150, 158
    Lactobacillus casei, 158
    Lactobacillus rhamnosus, 158
    Lactobacillus sake, 158
    Lactococcus lactis, 158
    yogurt bacteria, 158
Lepidoptera, 34
    butterflies, 34
    mites, 34
    moths, 34
Leucine rich repeats (LRRs), 63
LG microwave oven, 4
Lignocellulosic fibers, 171
Linseed oil (LO), 3
Liquid-nitrogen cooled deep depletion, 137
Literature scanning, 3
Lithography techniques, 137
Little plastic deformations, 112
Local electromagnetic field, 136
Localized surface plasmon polariton reso-
    nances, 136
    plasmon resonance, 136
    simply Plasmon, 136
Long chain biomolecules, 56
Low infectious dose, 50, 53, 54
Lowest unoccupied molecular orbital
    (LUMO), 89

M

Macroalgal EPS, 147
    industrial scale, 147
    product recovery, 147
    rapid growth, 147
    seasonal variations, 147
Magic mold releasing agent, 173

Mammalian cells, 19
Matrix reservoir technology, 32
Mature macrogametes, 62
Mechanical integrity, 17–20, 34
Mechanical properties, 17, 20, 23,
    28–30, 32, 34, 35, 37–41, 67, 124, 126,
    170–172, 177, 181, 186, 192
Mechanical testing, 28, 133
Medical expenses, 53
Melt recrystallization model, 109
Mesh-like structures, 66
    G. lamblia cyst walls, 66
Metal alkoxides, 3
Metal nanoparticles, 3, 95, 136, 197
Metal structure acts, 136
    antenna, 136
Metallic nanostructures, 136
Methacrylate (MA), 30
Microbial exopolymers, 60
Microbial origin, 60
Microcrystalline cellulose, 102–104, 172
Microfibrillated cellulose (MFC), 103
Microscope slide, 58, 61
Microwave (MW), 2, 3
Middle layer (media), 16
    collagen type I, 16
    elastin, 16
    proteoglycans, 16
    smooth muscle cells (SMCs), 16
Monitoring methods, 59
Morphological studies, 4
    transmission electron microscopy
    (TEM), 4
morphology of composites, 79
    do pants, 79
    solvents, 79
Mueller-Hinton agar plates, 10
Multi lamellar structure, 25
Multiscale porosity, 74

N

Nanocomposite, 2, 74, 77, 78, 80, 82, 84,
    93–95, 116, 117
Nanoparticles, 3, 8, 74, 76–95, 136
Nanosized cellulose, 103
    cellulose nanofibers (CNF), 103

cellulose nanowhiskers (CNW), 103
Native saphenous vein, 38
    dynamic compliance, 38
    burst pressure data, 38
Native vessel, 16, 17, 20
    external/internal thoracic arteries, 16
    radial arteries, 16
    saphenous veins, 16
Natural fiber reinforced polymer compos-
    ites, 124
    electrical loadings, 124
    mechanical, 124
    thermal, 124
Natural scaffolds, 17
    collagen, 17
    elastin, 17
    fibronectin, 17
Natural synthetic materials, 171
    amorphous silica gel, 171
    chitosan, 171
    fruit peel, 171
    crystalline lamellar talc-like, 171
    clay-like, 171
    inorganic phosphate, 171
Nature of the surface, 55
    charge and hydrophobicity, 55
Neo-tissue matures, 39
Noble metal nanoparticles (NPs), 136
Noncontact mode, 137
Nonhydrotrope acids, 80
    HCl, 80
    p-dodecylbenzenesulfonic acid
    (DBSA), 80
Nonmineralised connective tissues, 155
    cartilage, 155
    gum, 155
    skin, 155
    tendon, 155
Nonthrombogenic behavior, 37
Non-woven silk fibrin mats, 35
Normal systolic blood pressure, 25
Novel biopolymeric flocculent, 62
Novel hybrid scaffolds, 30
Nucleic acid targets, 59

O

Ocular drug pilocarpine, 156
(OO)cysts-surface association, 62
   interaction, 55
     (oo)cysts treatment, 55
      formalin, 55
Ophthalmic drug delivery system, 156
Ophthalmic drug echothiophate iodide, 156
   treatment of glaucoma, 156
Optical properties, 74, 76, 136, 140, 142
Oral administration, 156
   pharmaceutical formulation, 156
     aqueous alginate, 156
     drug (paracetamol) calcium ions, 156
     sodium citrate forms gel, 156
Organic matrix, 3
   alkyds, 3
   epoxies, 3
   polyols, 3
   polyester amides, 3
   polyurethanes, 3
Organic polymer shortcomings, 77
   inadequate mechanical strength, 77
Organic-inorganic hybrids, 3, 12
Oxidative polymerization, 74, 76, 82, 95
Oxygen permeability, 176, 187, 192
Oxygen transmission rate (OTR), 176

P

Para-film sealed plates, 10
Parameters, 18, 24, 86, 117, 138, 139, 181
   system parameters, 18
     molecular weight, 18
     molecular weight distribution, 18
     polymer solution viscosity, 18
     conductivity, 18
   process parameters, 18
     electric potential, 18
     flow rate, 18
     temperature, 18
     humidity, 18
Parasitic protozoans, 56, 59

Parasite developmental cycle, 62
Pathogen distribution, 59
Pathogen, 50, 52–54, 57, 59, 66
   acute self-limiting gastroenteritis, 52, 54
   cryptosporidiosis, 50, 52, 53
Persistent diarrhea, 52
   Cryptosporidium, 52
Petro-based counterparts, 2
Petroleum-based plastics, 170
Pharmaceutical applications, 147
Photocatalysis, 95, 198
Phromium tenax, 124
Physical removal methods, 56
   chemically assisted filtration, 56
   coagulation, 56
   flocculation, 56
PLA applications, 102
   drug delivery, 102
   medical, 102
   textile or packaging applications, 102
PLA based films, 102
   tensile testing, 102
   thermo gravimetric analysis (TGA), 102
   wide angle X-ray diffraction (WAXD), 102
PLA properties, 102
   barrier, 102
   mechanical, 102
   thermal, 102
Plant oils (PO), 2
   cost effective renewable resources, 2
   domestically abundant, 2
   environment friendly, 2
Plaque buildup, 16
   disease, 16
   hindered blood flow, 16
   inflammation, 16
   injury, 16
Plasma treatment, 22
Plasmon resonance, 76, 136–142
Plastic fiber composites, 124
   big blue stem, 124
   distillers dried grain, 124
   pine wood, 124
   polypropylene, 124

soybean hulls, 124
PO Polymers, 2
    lack properties of rigidity, 2
    strength, 2
    thermal stability, 2
Poly (Lactic Acid), 102, 103, 108–110,
    117, 170, 171
Poly(ethylene glycol) (PEG), 103
Polyacrylates, 64
    Dimethylacrylamide (DMAA), 64
    Diethylacrylamide (DEAA), 64
Polyester composites, 125–128, 132, 133
Polyglecaprone (PGC), 38
Polyglyconate, 38
    surgical suture materials, 38
        biocompatible-degradation rate, 38
        flexible, 38
Polymer microarrays, 61
    cell types, 61
        bacteria, 61
        human embryonic cells, 61
        human renal tubular epithelial cells,
        61
        human skeletal progenitor cells, 61
        mouse bone marrow dendritic cells,
        61
        suspension cells, 61
Polymer science, 2
Polymer surface properties, 62
    polymer composition, 62
    surface roughness, 62
    wettability or hydrophobicity, 62
Polymerase chain reaction (PCR), 59
Polymeric materials diverse, 50
Polymer-rich outer layer, 40
Polymers, 2, 17, 18, 25–30, 37, 39, 40,
    50–52, 55, 59–63, 65–67
Polymethylmethacrylate (PMMA), 176
Polymorphs of TiO2, 77
    anatase (tetragonal), 77
    brookite (orthorhombic), 77
    rutile (tetragonal), 77
Polynesian microbial mats, 157
Polyol, 2–12
Polyol backbone, 5, 6, 8

Polypeptide chains, 19, 30
    glycine, 19
    hydroxyl proline, 19
    proline, 19
Polypyrrole, 74, 78, 82, 84, 95
    batteries, 74
    photocatalysts, 74
    photovoltaic devices, 74
    supercapacitors, 74
Polypyrrolidone, 170, 174, 177, 191, 192
Polysaccharides, 146, 147, 154, 158, 159
    Levans, 146
    Dextrans, 146
Polyvinyl alcohol, 172, 192
Porous structure, 80, 81, 91
    photo catalytic, 81
    photovoltaic applications, 81
Postingestion trophozoite stage, 51
Potent antioxidant activities, 150
    reactive oxygen species, 150
        hydrogen peroxide, 150
        hydroxyl radical, 150
        inhibit lipid peroxidation in vitro,
        150
        superoxide radical, 150
Predict interactive phenomenon, 62
Preliminary investigative studies, 2
    antibacterial behavior, 2
    antibacterial studies, 5
    morphology, 2
    structure, 2
    water solubility, 2
Premature hydrolysis, 17
Preparation methods for PPy, 76
    concentrated emulsions, 76
    photochemistry, 76
    plasma, 76
    radiolysis, 76
Probiotic treated group, 154
    wound healing, 154
        fibrobast proliferation, 154
        initial macrophage infiltration, 154
Progenitor cells, 40
Protein synthesis inhibitory antibiotics,
    159
    amino acid deprivation, 159

chloramphenicol, 159
Protein-protein interactions, 63
Protozoal cysts, 55, 60, 61
Protozoal pathogens, 50
  Cryptosporidium, 50
  Giardia spp, 50
Pure silk fibroin, 36
Pyrene, 137, 138, 140–143
Pyrrole ring vibration, 86

Q

Quantitative microbial risk assessment
  (QMRA), 56
Quartz substrates, 65

R

Raman scattering, 136
Rat aorta, 35
  end to end anastomosis model, 35
Reducing agents, 200
Regular monitoring, 57
Reinforcement, 3, 117, 124, 127, 133, 170,
  171, 186
Ribbon like structure, 126
  coated waxy, 126
  materials on filaments, 126

S

Sample's morphology, 79
Sand cyst interactions, 55
  cell surface hydrophobicity, 55
  functionality, 55
  surface chemical charge, 55
Saphenous vein, 16, 20, 38
Scanning electron microscopy (SEM), 34,
  79, 102, 106, 177
Scar-tissue fibrosis, 19
Secondary metabolites, 146, 160
SERS, 136–142
Shear-stress resistance, 20
Shenzhen Bright China Industrial Co., 104
Short shelf-life products, 171
  blister packages, 171
  containers, 171
  drinking cups, 171
  overwrap and lamination films, 171

salad cups, 171
sundae, 171
Signaling pathway, 35
Signal-to-noise ratio (SNR), 140
Signs of thrombosis, 33
Silicone sleeves, 18
Silk fibroin scaffold, 35, 36
Silylated hydroxypropylmethycellulose
  (sHPMC), 155
Size-dependent properties, 76
  optical properties, 76
  quantum confinement in semiconductor
    particles, 76
  super para magnetism, 76
  surface plasmon resonance, 76
Small diameter vascular graft, 32, 33, 39,
  40
Small intestinal sin mucosa (SIS), 18
SNR, 140–142
Sodium methoxide, 4, 5
Softer biomatrix-rich, 40
Standard chlorination disinfection proce-
  dures, 50, 52
Stems of turmeric, 125
  hard Bunwar rubber sheets, 125
Stereocomplex crystal form, 106
Steric interaction, 56
  carbohydrates, 56
  proteins, 56
Stress-strain curves, 112
Stresswhitening, 112
Structure-activity relationships, 51
Substantial splaying, 24
Sulfated linear polysaccharides, 153
  heparin sulfate, 153
    blood coagulation cascade, 153
Sulfated silk fibers, 36
  standard electrospinning method, 36
    chlorosulfonic acid, 36
    pyridine, 36
Surface characterization, 38
  cell attachment, 38
  proliferation, 38
  regeneration, 38
Surface functional groups, 18
  protein, 18

cell attachment, 18
Surface-enhanced Raman scattering
    (SERS), 136
Symmetric ring stretching, 86
Synthetic vascular engineering, 34

T

Tailor-made SERS substrates, 136
Tensile energy absorption, 112, 114
Tensile testing, 24, 102
Tetraethoxy orthosilane (TEOS), 2
TGA diagrams, 87
Therapeutic applications, 148, 159
    human health, 148
    medicine, 148
Thermoplastic biodegradable polymers,
    171
    poly (lactic acid) (PLA), 171
    polyhydroxyalkanoate (PHA), 171
    polycaprolactones (PCL), 171
Timely vigilant observation, 5
TiO_2, 82, 86
Transmission electron microscopy (TEM),
    4, 74, 79, 84, 105, 108
Tri-layered electrospun scaffolds, 37
Triple helix structure, 19
Tropoelastin fibers, 24
Tubular conduit, 20, 38
Tubular mandrel, 24
Tubular morphology, 80
Tumor necrotic factors (TNF), 151
Tunable structural properties, 34
Turmeric, 124–133
    golden spice, 125
    medicinal plant, 125
    spice of life, 125
    time immemorial, 125
Turmeric extracted fiber, 125
    turmeric stem fiber (TSF), 125
    turmeric petiole fiber (TPF), 125
Turmeric fiber, 124–133

U

Ultimate elongation, 26
Ultra high vacuum set up, 137
Ultra-fine grade powder, 82

Ultrasonic bath, 4, 93, 104
Ultrasonic treatment, 102, 104, 108, 109
Ultra-thin polymer micronano fibers, 18
Uniform loan of culture, 10
Uniform morphology, 190
Uniform pore size, 148
Uniform wall-thickness, 20
Untreated turmeric stem, 130
    petiole fiber, 130
    impact loading, 130
    impact strength, 130
UV-vis absorption spectra, 79, 88, 89

V

Vascular endothelial growth factor
    (VEGF), 31
Vascular grafts, 16, 19, 21, 24, 29, 32–34,
    37, 39, 40
Vascular proteins, 37
    collagen, 37
    elastin, 37
    gelatin, 37
Vascular regeneration, 36, 40
Vascular scaffolds, 17, 18, 39
    cell migration, 17
    proliferation, 17
Vascular tissue engineering, 16, 18, 19, 22,
    26, 30, 31, 36, 37, 39, 40
Vascular tissue graft, 20, 41
Vascular tissue regeneration, 31, 40
Vascular tissue scaffolds, 17
    nature and origin, 17
        bio-hybrid, 17
        natural, 17
        synthetic, 17
Vegetable Oils, 12
Versatile applications, 2
    adhesives, 2
    antimicrobial agents, 2
    coatings, 2
    lubricants, 2
    paints, 2
Vibrational spectroscopic technique, 136
Virtanen data, 64
Visible gray region, 8
Volmer-weber growth, 137

W

Water protozoa, 67
Water quality monitoring, 67
Water treatment plants, 53
Water vapor permeability, 176, 189, 192
Waterborne pathogens, 67
Waterborne protozoa, 52, 58, 61, 65–67
    cyclospora, 52
    entamoeba, 52
    toxoplasma, 52
Waxy materials, 126
Wide-angle X-ray (WAXD), 105

X

Xe-arc lamp, 137

Y

Yields nanotube composites, 80

Z

Zone of Inhibition, 10, 11
zoonotic species, 53
    acute neonatal diarrhea, 53
      livestock, 53
2-(trifluromethyl)benzoylchloride, 172,
    173, 191, 192

9 781774 633496